REVISE EDEXCEL GCSE (9–1)
Physics

REVISION WORKBOOK

Higher

Series Consultant: Harry Smith
Author: Catherine Wilson

A note from the publisher

In order to ensure that this resource offers high-quality support for the associated Pearson qualification, it has been through a review process by the awarding body. This process confirms that this resource fully covers the teaching and learning content of the specification or part of a specification at which it is aimed. It also confirms that it demonstrates an appropriate balance between the development of subject skills, knowledge and understanding, in addition to preparation for assessment.

Endorsement does not cover any guidance on assessment activities or processes (e.g. practice questions or advice on how to answer assessment questions), included in the resource nor does it prescribe any particular approach to the teaching or delivery of a related course.

While the publishers have made every attempt to ensure that advice on the qualification and its assessment is accurate, the official specification and associated assessment guidance materials are the only authoritative source of information and should always be referred to for definitive guidance.

Pearson examiners have not contributed to any sections in this resource relevant to examination papers for which they have responsibility.

Examiners will not use endorsed resources as a source of material for any assessment set by Pearson.

Endorsement of a resource does not mean that the resource is required to achieve this Pearson qualification, nor does it mean that it is the only suitable material available to support the qualification, and any resource lists produced by the awarding body shall include this and other appropriate resources.

For the full range of Pearson revision titles across KS2, KS3, GCSE, Functional Skills, AS/A Level and BTEC visit:
www.pearsonschools.co.uk/revise

Contents

Edexcel publishes Sample Assessment Material and the Specification on its website. This is the official content and this book should be used in conjunction with it. The questions in *Now try this* have been written to help you practise every topic in the book. Remember: the real exam questions may not look like this.

Key concepts

1 Complete the table for units of physical quantities and their abbreviations.

Guided

ampere	....	joule	....	pascal	Pa	coulomb	....
mole	....	watt	....	newton	....	ohm	....

(2 marks)

2 Explain the difference between a base unit and a derived unit.

....

.... **(2 marks)**

3 Convert each quantity.

(a) 750 grams to kilograms

> 1000 g = 1 kg; 1000 W = 1 kW; 60 s = 1 minute; 1000 mm = 1 m; 1 000 000 J = 1 MJ

.... kg **(1 mark)**

(b) 0.75 kilowatts to watts

.... W **(1 mark)**

(c) 25 minutes to seconds

.... s **(1 mark)**

(d) 30 millimetres to metres

.... m **(1 mark)**

(e) 3 megajoules to joules

.... J **(1 mark)**

4 Write each quantity in the unit shown and then in standard form.

(a) frequency of 2.5 kHz

Guided

Hz: 2.5 kHz = 2500 Hz

standard form: 2.5×10^3 Hz **(2 marks)**

(b) length of 8 nm

m: ....

standard form: .... **(2 marks)**

5 Calculate the speed of a car that takes 10.5 s to travel 75 m. Give your answer to 5 significant figures.

> Look at the number that follows the significant figure you are asked to consider (in this case the 5th one). If it is greater than 5, round the 5th figure up, if it is less then round down, e.g. 1.23076923 would become 1.2308 to 5 significant figures.

speed = ... m/s **(2 marks)**

Scalars and vectors

1 (a) Write each quantity in the correct part of the table.

acceleration displacement speed energy
temperature mass force velocity
momentum distance

A scalar has only a magnitude (size) but a vector has both a magnitude **and** a direction.

Scalars	Vectors
energy distance mass speed temperature	force velocity displacement momentum acceleration

(2 marks)

(b) (i) Give one example of a scalar from your table and explain why it is a scalar.

Guided

Speed is a scalar because it has size but no specific direction **(2 marks)**

(ii) Give one example of a vector from your table and explain why it is a vector.

Acceleration is a vector because it has size

and specific direction **(2 marks)**

2 At the swimming pool, two swimmers are practising for a swimming gala. They swim from opposite ends of the pool. The first swimmer dives in from the left side and swims the length of the pool at a velocity of 1.3 m/s. The second swimmer then swims from the right at a velocity of –1.4 m/s.

(a) Explain why the velocity is used in this example instead of the speed.

Because there is a stated direction of the way they are swimming which velocity as a vector **(2 marks)**

(b) Explain why the second swimmer's velocity has a negative value.

As there are in the opposite direction **(1 mark)**

3 (a) Which of the following is not a scalar?

☐ **A** energy

☐ **B** temperature

☐ **C** mass

☒ **D** weight **(1 mark)**

(b) Give a reason for your answer to (a).

As it has both size and direction **(1 mark)**

4 An aeroplane flies in a straight line between two airports. The pilot knows that there will be a strong wind blowing at an angle of 60° to the direction in which the aeroplane will be travelling. Explain why it is important that the pilot uses vectors when planning the route the aeroplane will take.

So he knows the direction

....

.... **(3 marks)**

Speed, distance and time

1 The distance/time graph shows a runner's journey from his home to the park.

(a) State the letter that corresponds to the part of the runner's journey where he:

(i) stops .... **(1 mark)**

(ii) runs fastest. .... **(1 mark)**

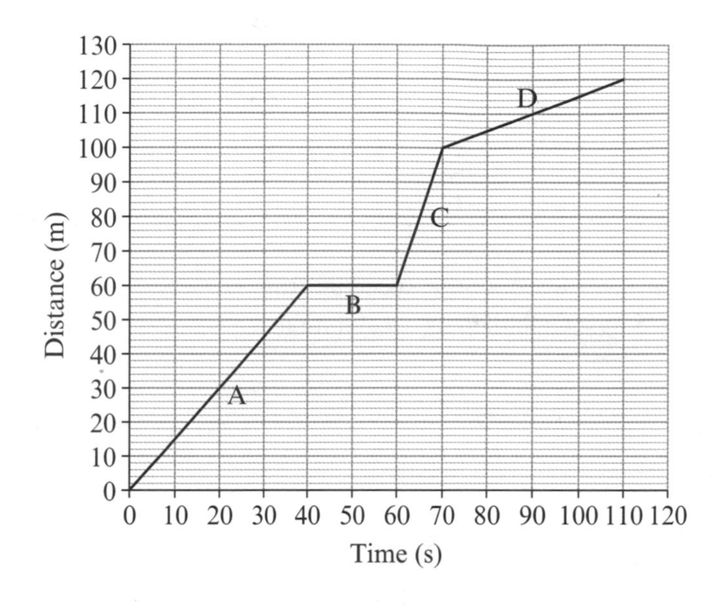

(b) Calculate the runner's speed in part A of his journey.

Guided

In part A, he travels .... m in .... s.

speed = distance ÷ ....

speed = ... m/s **(3 marks)**

(c) When the runner arrives at the park his displacement from home is less than the distance he has travelled. Explain this difference.

....

.... **(2 marks)**

2 The lift in a wind turbine tower takes 24 s to go from the ground to the generator 84 m above.

> Speed = distance ÷ time only.
> Velocity is speed in a given direction.
> For example, speed = 20 m/s but velocity = 20 m/s East

(a) Calculate the speed of the lift. State the unit.

speed = ... unit .... **(3 marks)**

(b) State the velocity of the lift.

.... **(1 mark)**

3 An athlete runs at a constant speed of 5 m/s around a running track. A complete lap is 400 m.

Calculate the time it takes for the athlete to complete one lap.

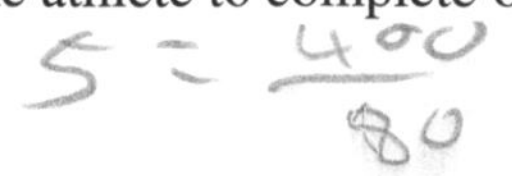

time = ... s **(2 marks)**

Had a go ☐ Nearly there ☐ Nailed it! ☐

Equations of motion

1 Draw a line from each symbol to its correct description. One has been done for you.

> Initial velocity means the velocity when time = 0.

Symbol	Description
v	acceleration
u	time
a	distance
x	final velocity
t	initial velocity

(2 marks)

2 (a) A racing car takes 8 seconds to speed up from 15 m/s to 25 m/s. Calculate its acceleration.

> You may find the equation $a = v - u \div t$ useful.

acceleration = ... m/s^2 **(3 marks)**

(b) The racing car now accelerates at the same rate for 12 seconds, from 25 m/s to a higher velocity. It travels 300 m during this time. Calculate its final velocity.

> You may find the equation $v^2 - u^2 = 2 \times a \times x$ useful.

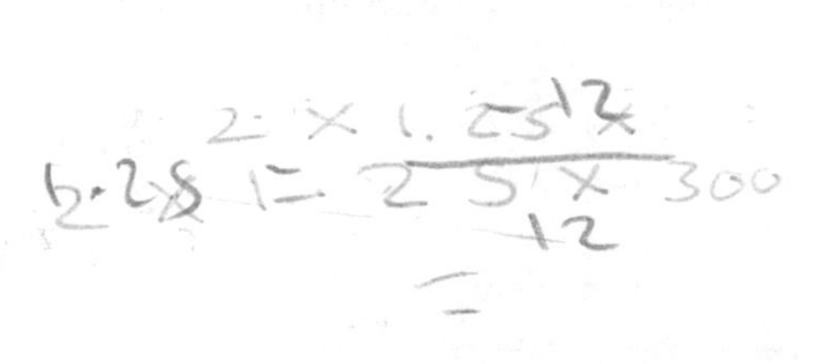

velocity = ... m/s **(3 marks)**

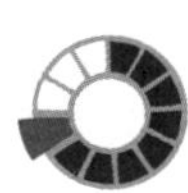

(c) The car now slows down to 5 m/s from the velocity calculated in (b) at a rate of $-2\ m/s^2$. Calculate how far the car travels when decelerating to this new final velocity.

distance = ... m **(3 marks)**

Velocity/time graphs

1 A cyclist takes 5 seconds to reach maximum velocity of 4 m/s, from being stationary, moving in a straight line.

(a) Calculate the cyclist's acceleration.

acceleration = ...unit .... **(3 marks)**

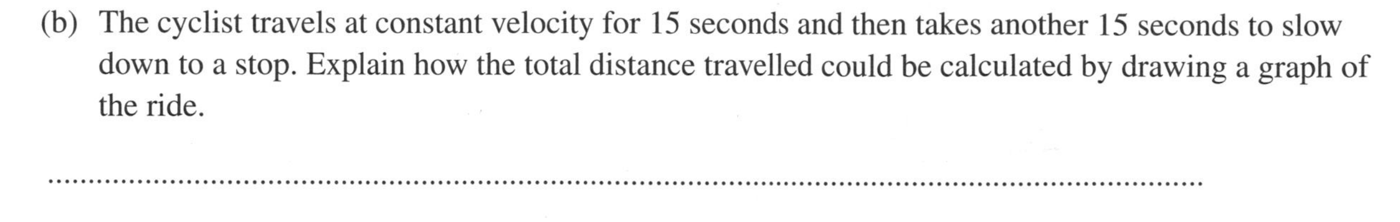

(b) The cyclist travels at constant velocity for 15 seconds and then takes another 15 seconds to slow down to a stop. Explain how the total distance travelled could be calculated by drawing a graph of the ride.

....

.... **(3 marks)**

2 The velocity/time graph shows how the velocity of a car changes with time.

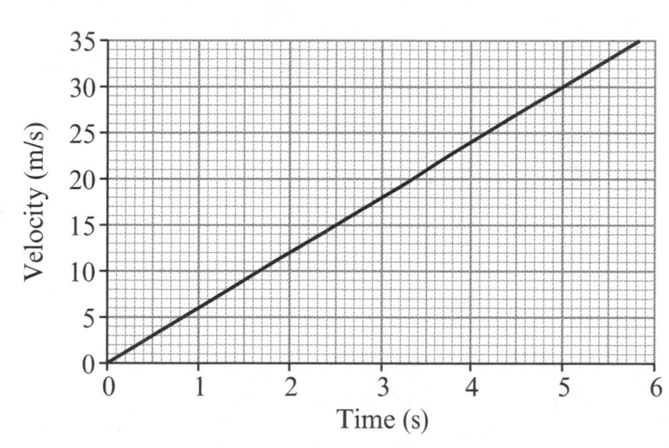

(a) This graph can be used to analyse the car's journey. Choose **two** correct statements that describe the information the graph shows.

☐ **A** the distance the car travelled

☐ **B** how long the car was stopped

☐ **C** the acceleration of the car

☐ **D** the constant velocity of the car

(2 marks)

(b) Draw a triangle on the graph to show the acceleration and the time taken. **(1 mark)**

(c) Calculate the acceleration of the car.

Guided

change in velocity = ... m/s, time taken for the change = ... s

acceleration = $\frac{\text{change in velocity}}{\text{time taken}}$ =

acceleration = ... m/s^2 **(2 marks)**

(d) Use the graph to calculate the distance travelled by the car in the first 5 s.

Work out the area under the graph.

distance = ... m **(2 marks)**

Had a go ☐ Nearly there ☐ Nailed it! ☐

Determining speed

1 Draw a line from each activity to its correct speed. One has been done for you.

Activity
commuter train
running
speed of sound in air
walking

Speed
330 m/s
1.5 m/s
3.0 m/s
55 m/s

(1 mark)

2 The diagram shows a light gate being used to measure the speed of a model vehicle.

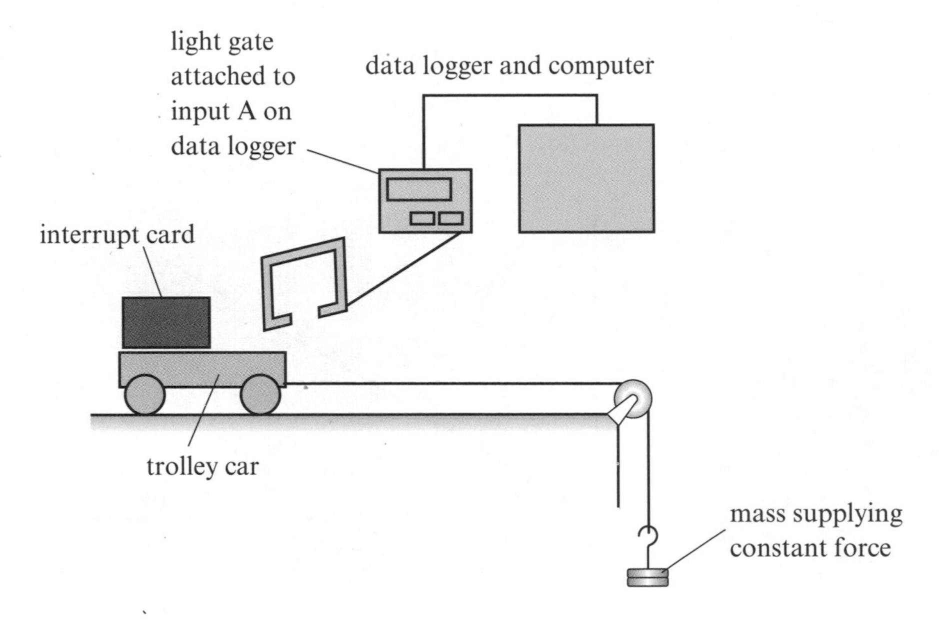

(a) Describe why a card is fixed to the vehicle in this experiment.

Guided

The light beam is ....

as it enters the light gate and this starts the timer. When the card has passed through, and ....

.... **(3 marks)**

(b) State how the speed is found using this method.

.... **(1 mark)**

3 State two reasons why using light gates and a computer may be a more reliable method than using a person with a stop watch and a ruler to measure the speed of a toy car.

> Light gates can measure instantly so computers can calculate speeds over very short distances. Consider this advantage over measurement by a person.

....

....

.... **(3 marks)**

Newton's first law

1 A submarine is travelling at a constant depth in the sea. It starts to move forwards. Draw a free-body force diagram for all the forces acting on the submarine. Label these forces. **(2 marks)**

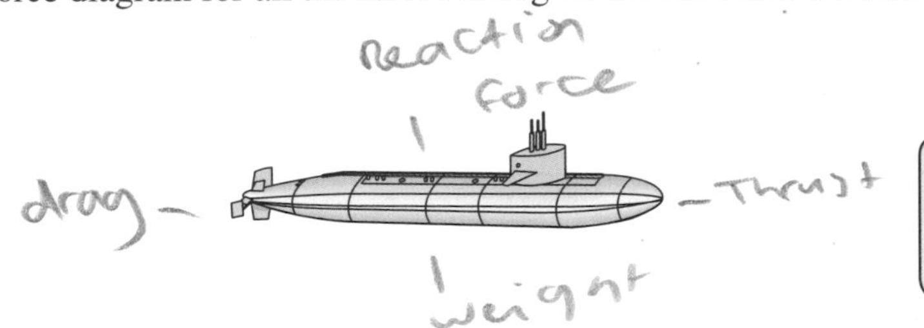

The lengths of the arrows on a free-body force diagram should be proportional to the sizes of the forces.

2 A speed skater is standing on the ice waiting for the start of a race.

(a) Describe the action and reaction forces acting on the skater and her skates.

Guided The action is the ......her pushing on the ice...... and the reaction is ......the ice pushing on her...... **(2 marks)**

(b) The race begins and the skater pushes against the ice producing a forward thrust on the skates of 30 N. There is resistance from the air of 10 N and friction on the blades of 1 N. Calculate the resultant force.

Add up all the forces in a straight line. Give forces that act opposite to the thrust a minus sign.

30 − 11 = 19

force = ...19...... N **(2 marks)**

(c) During the race the resistive forces become equal to the forward thrust. Explain what happens to the velocity of the skater.

......It will become constant......

......and in a straight line...... **(2 marks)**

(d) At the end of the race the skater stops skating. Explain what happens next before the skater comes to a halt.

....

.... **(2 marks)**

3 A space probe falls towards the Moon. In the Moon's gravitational field the probe has a weight of 1700 N. The probe fires rockets giving an upward thrust of 1900 N.

(a) Calculate the resultant force on the space probe.

1900 − 1700

resultant force = ...200...... N **(2 marks)**

(b) Explain the changes in the probe's velocity.

......It decreases......

.... **(2 marks)**

Newton's second law

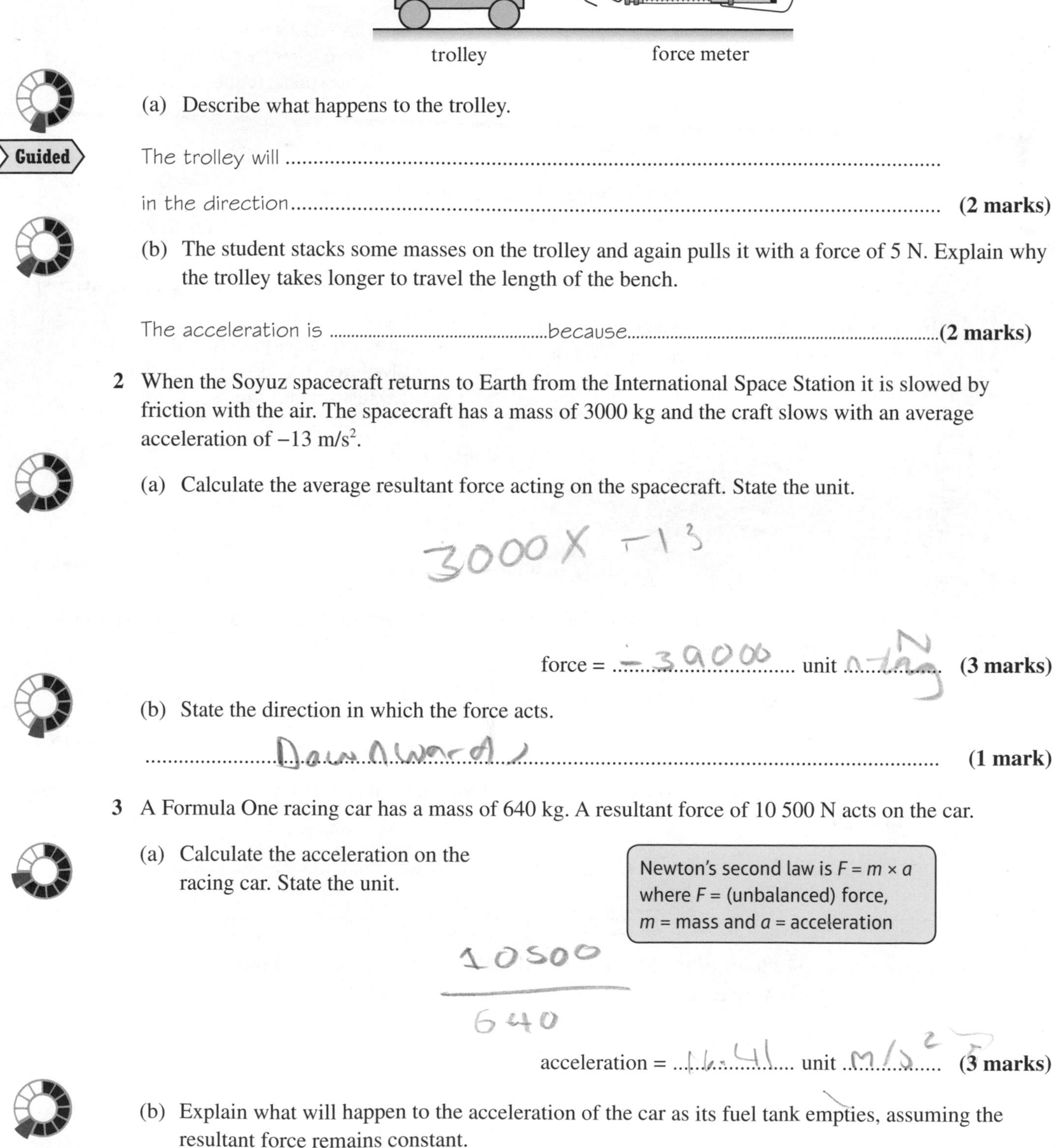

1 In an experiment a student pulls a force meter attached to a trolley along a bench. The trolley has frictionless wheels. The force meter gives a reading of 5 N.

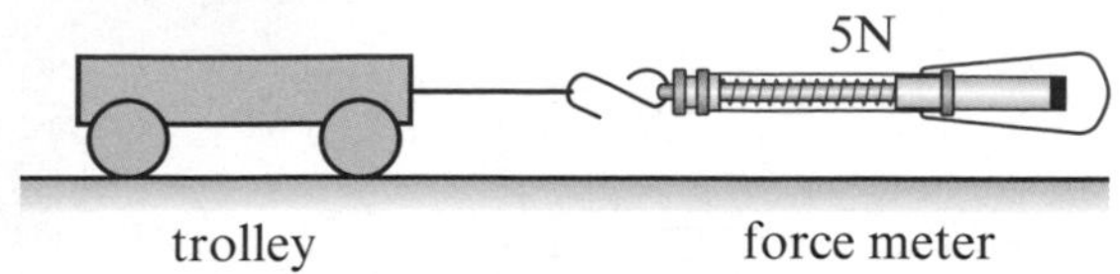

(a) Describe what happens to the trolley.

Guided

The trolley will ....

in the direction .... **(2 marks)**

(b) The student stacks some masses on the trolley and again pulls it with a force of 5 N. Explain why the trolley takes longer to travel the length of the bench.

The acceleration is .... because.................................... **(2 marks)**

2 When the Soyuz spacecraft returns to Earth from the International Space Station it is slowed by friction with the air. The spacecraft has a mass of 3000 kg and the craft slows with an average acceleration of –13 m/s².

(a) Calculate the average resultant force acting on the spacecraft. State the unit.

force = ... unit .... **(3 marks)**

(b) State the direction in which the force acts.

.... **(1 mark)**

3 A Formula One racing car has a mass of 640 kg. A resultant force of 10 500 N acts on the car.

(a) Calculate the acceleration on the racing car. State the unit.

> Newton's second law is $F = m \times a$ where F = (unbalanced) force, m = mass and a = acceleration

acceleration = ... unit .... **(3 marks)**

(b) Explain what will happen to the acceleration of the car as its fuel tank empties, assuming the resultant force remains constant.

....

.... **(2 marks)**

Weight and mass

1 The lunar roving vehicle (LRV), driven by astronauts on the Moon, has a mass of 210 kg on Earth. State the mass of the unchanged LRV on the Moon. Give a reason for your answer.

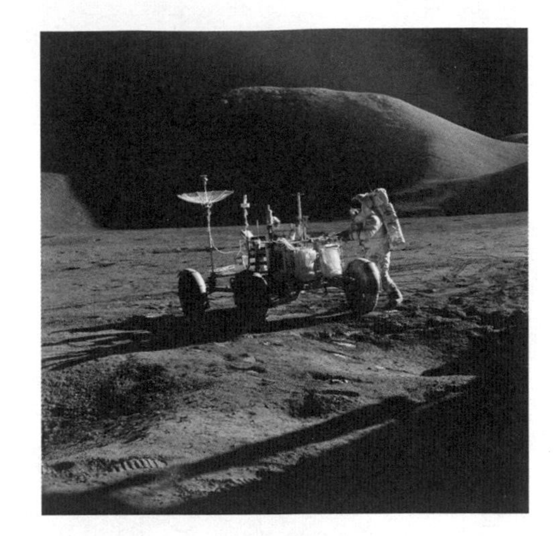

Guided

The mass of the LRV on the Moon is ..210.... kg

because ....

....

.... **(2 marks)**

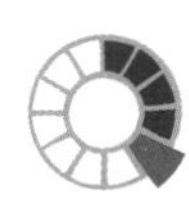

2 Which of the following is **not** a description of weight?

☐ **A** Weight is a type of force.

☒ **B** Weight is measured in kilograms (kg).

☒ **C** The weight of a mass changes according to gravitational field strength.

☐ **D** Weight is measured in newtons (N). **(1 mark)**

3 Calculate the total weight of a backpack of mass 1 kg, containing books with a mass of 2 kg and trainers with a mass of 1.5 kg. Take gravitational field strength (g) to be 10 N/kg.

Use the equation relating weight to mass and gravitational field strength.

weight = ...4.5.... N **(3 marks)**

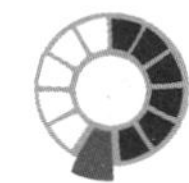

4 Kate is about to fly to Europe on holiday. The total baggage allowance is 20 kg. Kate only has scales that weigh in newtons. Determine the items that Kate can take on holiday, as well as her clothes, to get the mass as close as possible to the baggage allowance. Show your calculations. Take gravitational field strength (g) to be 10 N/kg.

laptop 45 N	camera bag 55 N	walking boots 25 N	jacket 35 N	clothes 105 N

total baggage = ...16.5.... kg **(3 marks)**

Force and acceleration

A ramp, a trolley, masses and electronic light gates can be used to investigate the relationship between force, mass and acceleration.

1 State one advantage of using electronic measuring equipment to determine acceleration compared to using a ruler and stopwatch.

....

....

.... **(2 marks)**

2 Describe the relationship between acceleration and mass.

.... **(1 mark)**

3 Explain why it is necessary to use two light gates when measuring acceleration in this experiment.

Guided

Acceleration is calculated by the change in speed ÷ time taken, so ....

....

.... **(2 marks)**

4 (a) Describe the conclusion that can be drawn from this experiment.

Guided

For a constant slope ....

....

.... **(2 marks)**

(b) Identify which of Newton's laws can be referred to in verifying the results of this experiment.

The quantities of force, mass and acceleration are linked in this equation.

.... **(1 mark)**

5 Suggest one hazard associated with this experiment and two safety precautions that could be taken to minimise the risk of harm to the scientist.

Consider the potential dangers of using accelerated masses or electrical equipment.

....

....

....

....

....

.... **(3 marks)**

Circular motion

Guided

1 (a) Explain why the velocity of a satellite is constantly changing even though its speed remains constant.

The velocity of an orbiting satellite changes because ....

even though ....

....

.... **(2 marks)**

(b) Explain why the Moon can be described as accelerating in its orbit round the Earth.

Refer to forces in your answer.

....

.... **(1 mark)**

2 What is the name of the force that is at 90° to the motion of a satellite?

☐ **A** acceleration

☐ **B** centripetal force

☐ **C** circular motion

☐ **D** orbiting force **(1 mark)**

3 Name the force that acts as the centripetal force in each example.
Add another example of your own for each force.

Force	Example 1	Example 2
	a lasso used to catch cattle	
	Venus orbiting the Sun	
	a cyclist going around a velodrome track	

(6 marks)

4 A student presents his project on the moons of Jupiter and uses a ball tied to string to model their motion by rotating the ball around in a circle, holding on to the string.

Consider how the forces act together.

(a) State the force that the string is modelling.

.... **(1 mark)**

(b) Explain why the ball is accelerating.

.... **(1 mark)**

Momentum and force

1 What is the momentum of a 10 000 kg lorry moving at 4 m/s?

☐ **A** 2500 kg m/s

☒ **B** 40 000 kg m/s

☐ **C** 14 000 kg m/s

☐ **D** 4×10^{-4} kg m/s **(1 mark)**

2 (a) Explain how force is related to momentum.

....

.... **(2 marks)**

(b) A car with a mass of 1500 kg is travelling at 25 m/s along a motorway. It crashes into a central barrier and stops in 1.8 seconds resulting in a momentum of zero. Calculate the change in momentum of the car.

1500 × 25 = 37 500

37 500 / 1.8

change in momentum = ... kg m/s **(3 marks)**

(c) Explain how a large force is exerted on a passenger in a vehicle in the event of a car crash and how this can be reduced.

Guided

The forces exerted on the passenger are large when ....

By fitting ....

So this will reduce the ....

....

....

.... **(4 marks)**

3 Calculate the force on a motorcycle of mass 500 kg as it speeds up from 10 m/s to 15 m/s in 20 s.

You may find this equation useful:
change in momentum = resultant force × time

force = ... N **(3 marks)**

4 Explain what a hockey player needs to consider when hitting the hockey ball with a hockey stick, to send the ball as far as possible down the pitch.

....

....

.... **(3 marks)**

Newton's third law

1 Select the statement that summarises Newton's third law.

- ☐ **A** For every action there is a constant reaction.
- ☐ **B** The action and reaction forces are different due to friction.
- ☐ **C** Reaction forces may be stationary or at constant speed.
- ☒ **D** For every action there is an equal and opposite reaction. **(1 mark)**

2 Calculate the momentum of a car with a mass of 1200 kg moving at 30 m/s from north to south.

> You may find this equation useful
> *momentum = mass × velocity*

Guided

momentum = 3600 ....

momentum = 3600 kg m/s in the South direction. **(3 marks)**

3 Dima and Sam are driving dodgem cars at a funfair. The total mass of Dima and his car is 900 kg. He is moving west at 1.5 m/s.

(a) Calculate the momentum of Dima and his car.

900 × 1.5

momentum = 1350 kg m/s **(1 mark)**

(b) Sam and his car also have a total mass of 900 kg but his car is travelling faster than Dima's car, at 3 m/s west. Sam's car collides with the back of Dima's car and both cars move forward together.

(i) Calculate the momentum of Sam and his car just before the collision.

900 × 3
= 2700

momentum = 2700 kg m/s **(1 mark)**

(ii) Explain what happens to the sum of the momentum of both cars after the collision.

They are both equal as the force exerted on them equal in opposite direction as momentum is so conserved **(2 marks)**

(iii) Calculate the velocity of both cars as they move off together after the collision.

~~1800~~ 2700 ÷ 900

velocity = 3 m/s **(3 marks)**

4 A skater with a mass of 50 kg skates across the ice at 7.2 m/s in a straight line travelling north. She collides with her stationary partner who has a mass of 70 kg. They glide off together northwards. Calculate the velocity with which the pair glide across the ice.

50 × 7.2 = 360 ÷ 120

velocity = 3 m/s **(3 marks)**

Had a go ☐ Nearly there ☐ Nailed it! ☐

Human reaction time

1 Reaction time is an important consideration in driving a vehicle safely. Which is the distance travelled due to the reaction time of a driver?

☐ **A** overall stopping distance

☒ **B** thinking distance

☐ **C** braking distance

☐ **D** reaction distance **(1 mark)**

2 Explain how human reaction time is related to the brain.

Guided

Human reaction time is the time between a stimulus occuring and a respo…

It is related to how fast quickly the brain **(2 marks)**

3 Explain how to measure human reaction times using a ruler.

Person A stand with there index finger and thumb opened to a gap of about 8cm. Then Person B holds a meter rule so that it **(3 marks)**

4 (a) State the range of reaction times of an average person to an external stimulus.

0.20 – 0.25 **(1 mark)**

(b) Describe why people in certain professions train themselves to improve their reaction times. Give two examples and comment on why improved reaction times would be important in each case.

Examples of professions you could use are driving instructors and helicopter pilots.

(4 marks)

5 A rabbit runs across the road 50 metres in front of a car. Calculate the reaction time of a driver who covers a distance of 25 metres travelling at a speed of 20 m/s between seeing the rabbit and putting his foot on the brake.

You may find this equation useful *speed = distance ÷ time*

Reaction time .... s **(2 marks)**

Stopping distance

1 (a) Write the word equation used to calculate overall stopping distance.

Thinking + breaking **(1 mark)**

(b) Calculate the overall stopping distance when a car increases its speed from 20 mph to 60 mph. Take thinking distance to be 6 m and braking distance to be 6 m when travelling at 20 mph.

(3 marks)

(c) Complete the table below to summarise the factors that affect overall stopping distance.

Separate the factors that may affect the reaction time of a driver from those that affect the vehicle.

Guided

Factors increasing overall stopping distance	
Thinking distance will increase if	**Braking distance will increase if**
The driver alcohol/drugs	the car's speed increases
the driver is distracted	Icy roads or wet
tired	Tyres broken

(2 marks)

(d) Compare the overall stopping distances of a car with worn tyres and a car with new tyres.

....

.... **(2 marks)**

2 Work is done on a moving car to bring it to rest. Calculate what force must be applied to the brakes of a car of mass 1500 kg travelling at 8 m/s for it to stop at the pedestrian crossing 75 m away.

force = ... N **(3 marks)**

3 Recent proposals have been made to increase the national speed limit in certain cases. Suggest how these proposals might increase the risk of damage to vehicles and their passengers.

Remember that kinetic energy is proportional to v^2.

....

....

....

.... **(3 marks)**

Had a go ☐ Nearly there ☐ Nailed it! ☐

Extended response – Motion and forces

A student investigates circular motion by tying a 57 g tennis ball to a string which is then rotated in a horizontal plane at constant speed. The student counts the number of rotations.

Explain how acceleration and centripetal force are considered in this experiment.

Your answer should include an example of how the experiment may be extended to improve data collection.

> You will be more successful in extended writing questions if you plan your answer before you start writing.
>
> The question asks you to give a detailed explanation of acceleration and centripetal force. Think about:
>
> - Why the tennis ball is described as accelerating.
> - How centripetal force is described.
> - How the student can improve his investigation by changing variables.
> - How the student can improve data collection.
> - Identify the importance of control variables.
>
> You should try to use the information given in the question.

....

....

....

....

....

....

....

....

....

....

....

....

....

....

....

.... **(6 marks)**

Energy stores and transfers

1 Which of the following is not an energy store?

☐ **A** chemical

☐ **B** light

☐ **C** thermal

☐ **D** kinetic

(1 mark)

2 Explain how an energy transfer diagram supports the law of conservation of energy.

Guided

The energy transfer diagram shows that ....

....

....

.... **(2 marks)**

3 A footballer has a breakfast of cereal and toast before setting off for a training session at the club. Complete the flow chart to show how energy is transferred to other stores.

Write the correct store of energy in each space.

....	→	....	→	....
energy in the breakfast		energy of the footballer		energy dissipated to the surroundings

(3 marks)

4 The bar graphs below illustrate energy stores before each energy transfer occurs. Add bars to the graphs to show energy stores for after each energy transfer has occurred.

(a) a bobsleigh at the top of a slope and halfway down the slope

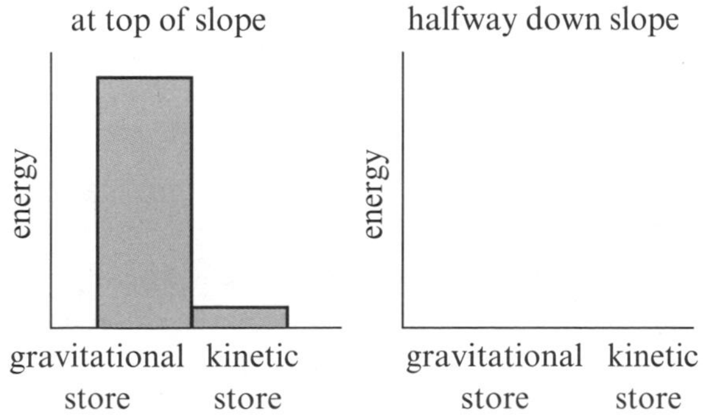

(b) a petrol lawnmower before use and in use

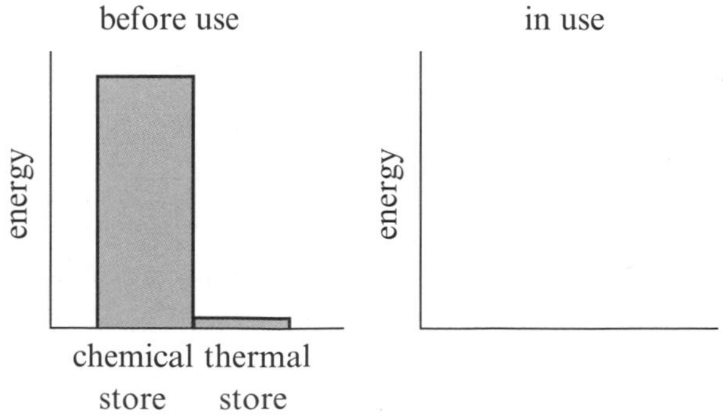

(3 marks)

(4 marks)

Efficient heat transfer

1 Identify the most suitable material, from the table below, for building an energy-efficient garage. Give a reason for your answer.

The larger the relative thermal conductivity, the more heat will be conducted through the material.

Material	Relative thermal conductivity
brick	1.06
concrete	1.00
sandstone	2.20
granite	2.75

....

.... **(2 marks)**

2 (a) Some houses are built with very thick walls. Explain how these walls help to keep the houses warm in the winter in cold countries.

....

....

.... **(2 marks)**

(b) In hot countries, such as Greece, traditional houses have thick walls with small windows. Explain why these houses in a hot country also have thick walls.

....

....

.... **(2 marks)**

3 A crane lifts a box to the top of a building. 1 000 000 joules is transferred to the gravitational store when the box is moved from the bottom to the top of the building. The crane uses fuel with 4 000 000 joules in a chemical store. Calculate the efficiency of the crane.

Guided

useful energy transferred = energy transferred to the box = ...

total energy used by the crane = the energy stored in the fuel = ...

efficiency = ... **(2 marks)**

4 (a) The motor in a food blender has an efficiency of 20%. The motor transfers 40 joules per second into the kinetic store. Calculate the energy that is transferred to the motor each second.

energy transferred each second = ... J **(3 marks)**

(b) State the power of the motor. Give the unit.

power = ... unit .... **(1 mark)**

Energy resources

1 Some of the sources of renewable energy listed below are only available at certain times, while other sources can be used at any time.

hydroelectric	tidal	solar	wind	geothermal

(a) Name the sources of renewable energy in the list that are always available.

Guided

Hydroelectric and .... **(1 mark)**

(b) Explain why it is an advantage to have a source of energy available at any time.

Think about how the weather affects some renewable energy sources.

Demand is greatest ....

Demand may be high when .... **(2 marks)**

2 A hydroelectric power station is used to produce electricity when demand is high.

(a) Explain why the hydroelectric power station is a reliable producer of electricity.

....

.... **(2 marks)**

(b) Give one reason why we cannot use hydroelectric power stations in more places in the UK.

....

.... **(1 mark)**

3 Comment on each statement referring to the use of fossil fuels with regard to environmental impact.

Think about the possible consequences of the statements describing the use of fossil fuels.

(a) Carbon dioxide is released as a result of burning fossil fuels.

....

.... **(2 marks)**

(b) Burning fossil fuels produces sulfur dioxide and nitrogen oxides.

....

.... **(2 marks)**

(c) Fossil fuel power stations can be built away from areas of natural beauty such as coasts, estuaries and mountains.

....

.... **(2 marks)**

4 Some people say that we have passed the time of 'peak oil'. After this time, the amount of crude oil extracted will decrease and prices for fuel will rise rapidly. Other people say that we will not pass this peak until 2020. Suggest why there is uncertainty about peak oil.

....

.... **(2 marks)**

Had a go ☐ Nearly there ☐ Nailed it! ☐

Patterns of energy use

1 The graphs show patterns of energy use and human population growth.

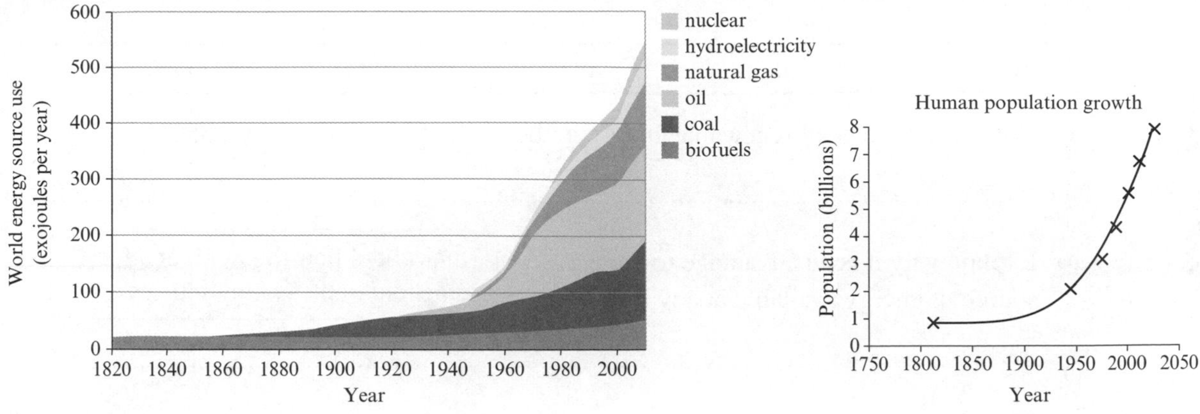

(a) Give three reasons why energy consumption rose significantly after the year 1900.

Guided

After 1900 the world's ....

There was development in ....

and .... **(3 marks)**

(b) (i) Identify which category of energy resources has been the main contributor to world energy consumption since the year 1900.

.... **(1 mark)**

(ii) Suggest two reasons why the consumption of energy resources has increased in the developed world.

....

.... **(2 marks)**

(iii) Suggest a reason why nuclear energy resources only appear after 1950.

.... **(1 mark)**

(iv) Identify a renewable resource from the graph that makes use of gravitational potential energy.

.... **(1 mark)**

2 If the patterns in energy consumption are similar to the patterns in the world's population growth, discuss the issues resulting from the continuing use of energy in the way shown in the graph in Q1.

Consider finite non-renewable resources and increasing demand due to population, transport and industrial growth.

....

....

....

....

.... **(6 marks)**

Potential and kinetic energy

1 Identify the correct equation for calculating gravitational potential energy.

☐ **A** $\Delta GPE = m \times v \times h$ ☐ **C** $\Delta GPE = m \times F \times a$

☐ **B** $\Delta GPE = ½ m \times v^2$ ☐ **D** $\Delta GPE = m \times g \times h$ **(1 mark)**

2 Calculate the kinetic energy of a cyclist and her bicycle, with combined mass of 70 kg, travelling at 6 m/s.

Guided Kinetic energy = ½ mv^2 so ....

so ....

kinetic energy = ... J **(2 marks)**

3 In the Middle Ages battering rams were used to smash down the doors of castles. The battering ram was a log of mass 2000 kg suspended by ropes. When the log was pulled back, it rose by 0.5 m.

0.5 m upwards

(a) Calculate the gravitational potential energy gained by the log when it was pulled back.

gravitational potential energy = ... J **(2 marks)**

(b) State how much work is done in pulling the log back.

work done = ... J **(1 mark)**

(c) Calculate the velocity of the battering ram as it reaches the bottom of the swing.

velocity = ... m/s **(4 marks)**

4 Explain the energy changes in a golf ball from when it is hit by the golfer to when it reaches the highest point in the air, in terms of gravitational potential energy and kinetic energy. Identify any energy losses and explain why they may occur.

....

....

....

....

....

.... **(4 marks)**

Had a go ☐ Nearly there ☐ Nailed it! ☐

Extended response – Conservation of energy

Millie plays on a swing in the park. The swing seat is initially pulled back by her friend to 30° to the vertical position and then released. Describe the energy changes in the motion of the swing.

Your answer should also explain, in terms of energy, why the swing eventually stops.

You will be more successful in extended writing questions if you plan your answer before you start writing.

The question asks you to give a detailed explanation of the energy changes as the swing moves backwards and forwards. Think about:

- How gravitational potential energy changes as the swing is pulled back.
- Points at where gravitational potential energy (GPE) and kinetic energy (KE) are at maximum and at 0.
- Where some energy may be lost from the system.
- Why the swing will eventually stop.

You should try to use the information given in the question.

....

....

....

....

....

....

....

....

....

....

....

....

....

....

....

.... **(6 marks)**

Waves

1 The table below lists statements about transverse and longitudinal waves. Identify which type of waves are described by writing T (transverse), L (longitudinal) or B (both) next to each statement.

Sound waves are this type of wave.	L	They have amplitude, wavelength and frequency.	B
All electromagnetic waves are this type of wave.	T	Seismic P waves are this type of wave.	L
Particles oscillate in the same direction as the wave.	L	They transfer energy.	B

(3 marks)

2 The diagram below shows a wave travelling through a medium.

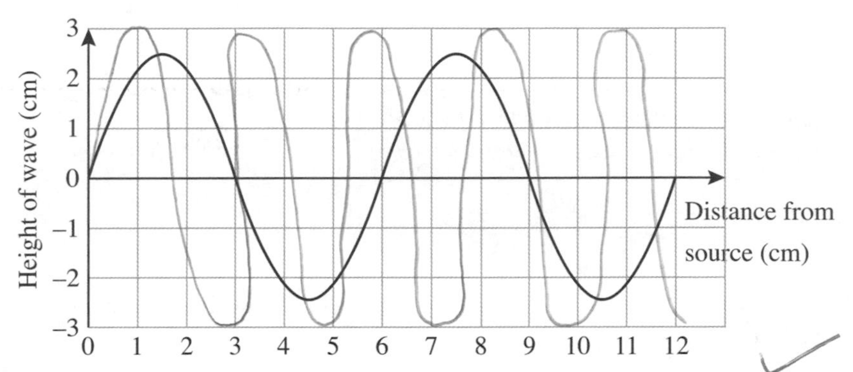

(a) What is the amplitude of the wave in the diagram?

☐ **A** 0.05 m ☑ **B** 0.025 m ☐ **C** 0.12 m ☐ **D** 0.10 m (1 mark)

(b) Determine the wavelength of the wave in the diagram.

You may find this equation useful $v = f \times \lambda$ or
wave speed = frequency × wavelength

V = 2 × 0.06
V = 0.12 m

wavelength = 0.12 × 0.06 m ~~(1 mark)~~

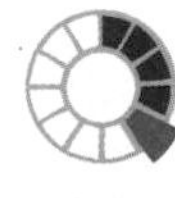

(c) Sketch a second wave on the diagram to show a higher amplitude and shorter wavelength. (2 marks)

3 When a wave travels through a material the average position of the particles of the material remains constant. Explain how this is correct for the types of waves found in:

(a) a sound wave travelling through the air

When a sound wave is generated each particle moves back and forth along the same direction that the sound is travelling (2 marks)

(b) ripples travelling across the surface of a pool.

When a water wave is generated the surface particles of water move in a direction at right angles to the direction the wave is travelling. (2 marks)

Wave equations

1 Whales communicate over long distances by sending sound waves through the oceans. It takes 20 seconds for the sound waves to travel in seawater between two whales 30 kilometres apart.

Calculate the speed of sound in water in metres/second. 30km = 30000 m

Guided Speed = Distance travelled by the waves (in metres) ÷ time taken (in seconds)

So speed = 30000 ÷ 20 = 1500 m/s

speed of sound = 1500 m/s **(3 marks)**

2 A sound wave has a wavelength of 0.017 m and a frequency of 20 000 Hz.

Calculate the speed of the wave in metres/second.

> You may find this equation useful $v = f \times \lambda$ or *wave speed = frequency × wavelength*

$v = 20000 \times 0.017$
$= 340$ m/s

wave speed = 340 m/s **(2 marks)**

3 An icicle is melting into a pool of water. Drops fall every half a second, producing small waves that travel across the water at 0.05 m/s. Calculate the wavelength of the small waves. State the units.

> Remember to write down the equation you are using before you substitute values.

$\lambda = \frac{v}{f} = 0.05$

$0.05 \times 2 =$ 0.025 m

wavelength = 0.1 unit m **(3 marks)**

4 A satellite sends signals to your TV using radio waves. It takes 0.12 s for the radio waves to travel from the satellite to your TV. The speed of light is 3×10^8 m/s. Calculate the distance of the satellite above the Earth in kilometres.

> You may find this equation useful $v = x \div t$ or *speed = distance ÷ time*

$x = \frac{v}{t} = \frac{3\times10^8}{0.12} = 2500000000$ m

$= 25000000$ km

$= 2.5 \times 10^7$ km

36 000 km

distance = 2.5×10^7 km **(3 marks)**

Measuring wave velocity

1 A tap is dripping into a bath. Three drops fall each second producing small waves that are 5 cm apart. Calculate the speed of the small waves across the water. State the units.

Guided

frequency of the waves (f) = ...

wavelength of the waves (λ) = ...

speed of waves = ... unit .... **(3 marks)**

2 How far apart are the crests of water waves with a frequency of 0.25 Hz travelling at a speed of 2 m/s?

Check your answers to Q1 to help you answer this question.

☐ **A** 0.5 m ☐ **C** 4 m

☐ **B** 0.125 m ☐ **D** 8 m **(1 mark)**

3 An oscilloscope screen shows a waveform.

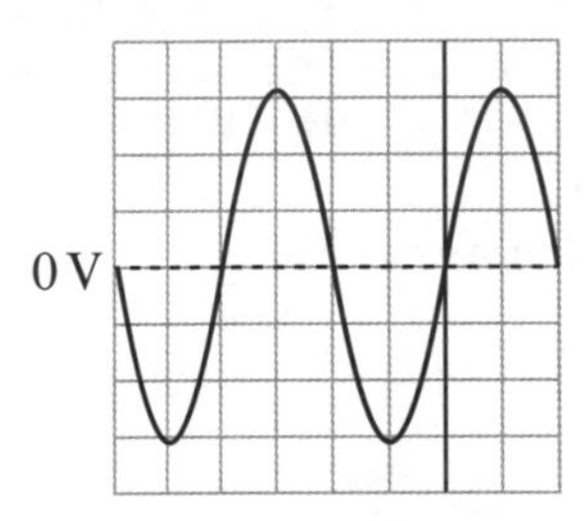

Each division in the horizontal direction is 5 ms. Calculate the frequency of the wave.

frequency = ... Hz **(3 marks)**

4 Eliza and Charlie set up an experiment to estimate the speed of sound in air. They use a brick wall at school and stand 50 metres away. Charlie knocks two pieces of wood together and Eliza measures the time of the echo using a stopwatch.

(a) Describe how Eliza and Charlie can use their measurements to calculate the speed of sound in air.

.... **(1 mark)**

(b) Suggest three changes that Eliza and Charlie could make to their experiment to make their results more reliable.

....

....

....

....

.... **(3 marks)**

Had a go ☐ Nearly there ☐ Nailed it! ☐

Waves and boundaries

1 Draw a line to match each wave phenomenon at boundaries with the correct explanation. One has been done for you.

Wave phenomena	Explanation
reflection	The wave energy is transferred into a thermal energy store.
refraction	The wave bounces back at a surface but does not pass through.
transmission	The wave energy is transferred.
absorption	The wave passes through but at a changed speed.

(2 marks)

2 (a) Which of the following will reflect the most sound energy in air?

☐ **A** curtain ☐ **C** stone wall

☐ **B** hedge ☐ **D** wooden fence **(1 mark)**

(b) Explain your answer to (a).

....

.... **(2 marks)**

3 Explain, in terms of particles, why sound travels faster in warm air than in cool air.

Sound waves are longitudinal (pressure) waves that travel by moving particles.

Guided

In warm air, the particles have more ....

....

....

....

.... **(3 marks)**

4 Discuss how the density of two materials affects the reflection and transmission of sound waves at the boundary. Your answer should include an example resulting in reflection or an example resulting in transmission.

Consider the how differences in densities of materials affects how a wave behaves at the boundary.

....

....

....

....

....

.... **(3 marks)**

Sound waves and the ear

1 Identify the correct sequence of sound waves in air when a guitar is played.

(a) Energy travels via longitudinal waves through the air.

(b) The air molecules near the guitar string vibrate in response to the movement of the string.

(c) The ear drum vibrates.

(d) Longitudinal waves are channelled into the ear canal.

☐ **A** (a), (c), (d), (b) ☐ **C** (c), (d), (b), (a)

☒ **B** (b), (a), (d), (c) ☐ **D** (a), (b), (c), (d) **(1 mark)**

2 Explain why a piano has strings of different lengths.

Strings of different lengths will vibrate at different frequencies.

Because they have different sound frequencies depending on thickness density size length, mass **(2 marks)**

3 Explain why sound travels more slowly in a gas than in a solid.

them as it can easily be compressed and are very compressible

(2 marks)

4 Some animals communicate using frequencies below 20 Hz. Others use frequencies over 20 kHz. Explain why humans cannot hear these sounds.

Guided

In the human ear, the ear drum will not .... if the frequency is

If there is no vibration ....

(3 marks)

Had a go ☐ Nearly there ☐ Nailed it! ☐

Ultrasound and infrasound

1 (a) State what is meant by the term infrasound.

Sound that is below 20hz and is used for discovering structure of earth **(1 mark)**

(b) Which of these sentences best describes infrasound?

☒ **A** Infrasound can only be detected when it travels through the ground.

☐ **B** Infrasound does not travel through the ground.

☐ **C** Infrasound travels further through the ground than 'normal' sound.

☐ **D** Volcanic eruptions do not produce 'normal' sounds. **(1 mark)**

(c) Infrasound detectors can be set up to record the direction that infrasound is coming from. Suggest how scientists use this to search for petroleum deposits in the Earth's crust.

Guided

Scientists set off .... and use ....

to receive the infrasound waves. These are then used to .... **(3 marks)**

2 Ultrasound can be used by oceanographers to explore the seabed. An ultrasound pulse travels through seawater at 1500 m/s and an echo is heard 4 seconds after transmission. Calculate the depth of the seabed at that point.

You will need to consider speed, distance and time, but be careful with distance.

1500 =

depth = ... m **(2 marks)**

3 Describe how the reflection of ultrasound is used to make a picture of a foetus in the womb.

....

....

....

.... **(4 marks)**

4 Ultrasound is used to produce images of the structure of computer chips as the waves are reflected by the different layers of materials. An ultrasound signal is sent into the top surface of a chip and an echo is detected from a layer in the chip after 0.5 nanoseconds. The speed of sound in the computer chip is 8400 m/s. Calculate the distance of the layer from the surface of the chip.

Maths skills

distance = ... m **(4 marks)**

5 Volcanologists can use infrasound produced during earth tremors to monitor seismic activity. Explain why this might have advantages for the general population.

....

....

.... **(3 marks)**

Sound wave calculations

1 Show that the wavelength of a wave with constant speed will halve if frequency is doubled. Refer to the wave equation in your answer.

Use the wave equation $v = f \times \lambda$ with numbers of your choice to show the relationship.

....

....

.... **(2 marks)**

2 The speed of sound varies in different materials.

(a) Describe how you would expect the speed of sound to change as it passes from a low-density material to a high-density material.

Guided

As sound waves pass from a low- to a high-density material the ....

.... **(1 mark)**

(b) Explain your answer to (a).

The denser the material ....

.... **(2 marks)**

Maths skills

3 The structure of the Earth can be analysed using information from seismic waves. It has been found that the speed of seismic waves increases as they go deeper into the Earth's semi-solid mantle.

(a) Suggest an explanation for this.

....

....

....

....

P-waves through the Earth

....

....

.... **(3 marks)**

(b) P-waves are refracted at various boundaries as they travel through the Earth. Explain what this tells us about the structure of the Earth.

....

.... **(2 marks)**

(c) Calculate the wavelength of seismic waves generated by an earthquake. The waves travel at 7 m/s with a frequency of 0.05 Hz.

wavelength = ... m **(2 marks)**

Had a go ☐ Nearly there ☐ Nailed it! ☐

Waves in fluids

1 A ripple tank is used to investigate waves.

(a) Describe how a ripple tank may be used to measure the frequency of water waves.

....

....

.... **(2 marks)**

(b) Describe how to find the wavelength of the waves in the ripple tank.

....

....

.... **(2 marks)**

(c) State the equation you can use with the data collected in (a) and (b) to determine wave speed.

.... **(1 mark)**

(d) Identify the control variable when using a ripple tank to investigate wave speed.

.... **(1 mark)**

2 Describe a suitable conclusion to the method of using the ripple tank in Q1. Your conclusion should include two factors that should be moderated in this experiment.

A ripple tank can be used to determine a value for ....

....

.... **(3 marks)**

3 The ripple tank experiment uses several pieces of equipment. Complete the table below to describe the hazard associated with each component and suggest a measure to minimise the risk of harm.

Identify the hazard and describe the safety measure for each mark.

Component	Hazard	Safety measure
water		
electricity		
strobe lamp		

(3 marks)

Extended response – Waves

A man has dropped his door key into the pool. The key appears to be in a different position because of the phenomena shown in the diagram.

Explain why the key appears to be in a different position.

Your answer should identify the actual position of the key.

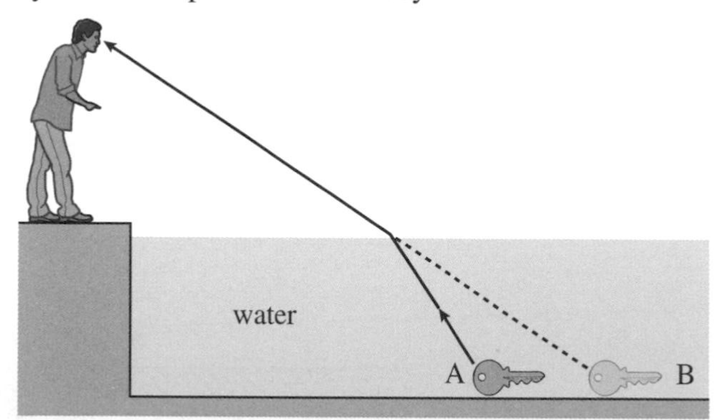

You will be more successful in extended writing questions if you plan your answer before you start writing.

The question asks you to give a detailed explanation of what happens when a light wave passes through the boundary between two different materials. Think about:

- The properties of waves that are illustrated.
- The types of materials involved in the wave behaviour shown in the diagram.
- The causes of the wave behaviour shown in the diagram.
- Why the man believes the key is in a different position from the actual position.

You should try to use the information given in the question.

....

....

....

....

....

....

....

....

....

....

....

....

....

....

.... **(6 marks)**

Reflection and refraction

1 State the law of reflection.

Guided

angle of.. = angle of .... **(1 mark)**

2 A ray of light is reflected from a plane mirror at an angle of 45°. Complete the diagram to show the incident ray, the normal, the reflected ray and the angles of incidence and reflection. **(4 marks)**

3 State what happens to a ray of light passing from a more dense material (water) into a less dense material (air).

Think about what happens when a light ray leaves a dense material.

.... **(1 mark)**

4 Light travelling through air hits a sheet of glass at an angle to the normal.

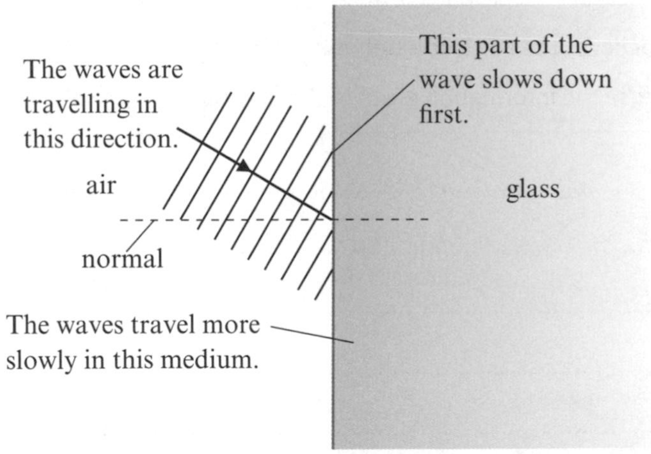

(a) Complete the diagram to show what happens as the light waves enter the glass. **(3 marks)**

The key points to show on the diagram are the change in direction and wavelength as the wave moves into the medium.

(b) Explain why refraction occurs whenever a wave crosses a boundary between materials of different refractive indices at angles other than normal to the boundary.

....

....

....

.... **(2 marks)**

Total internal reflection

1 (a) When light passes from one material to another, state what causes it to:

(i) bend towards the normal

.... **(2 marks)**

(ii) bend away from the normal.

.... **(2 marks)**

(b) State when light does not change direction when passing from air to glass.

Guided When light enters a glass block at .... **(1 mark)**

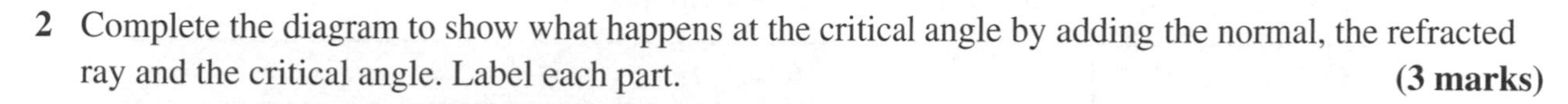

2 Complete the diagram to show what happens at the critical angle by adding the normal, the refracted ray and the critical angle. Label each part. **(3 marks)**

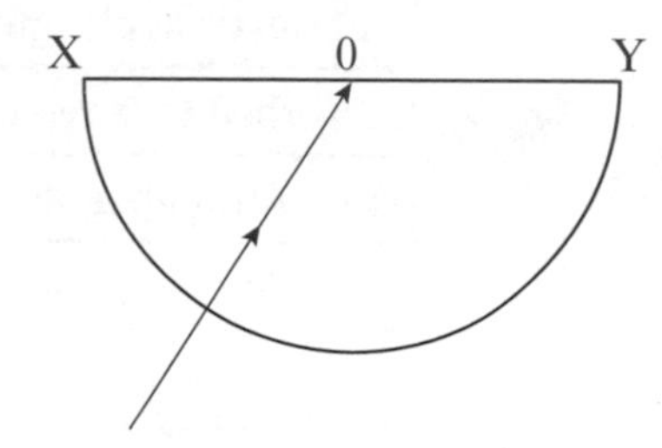

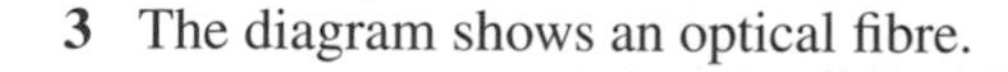

3 The diagram shows an optical fibre.

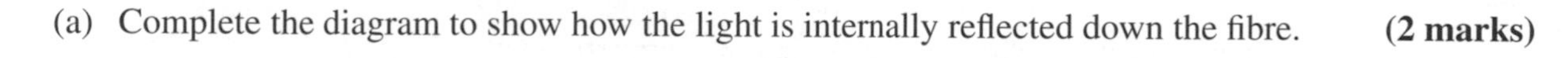

(a) Complete the diagram to show how the light is internally reflected down the fibre. **(2 marks)**

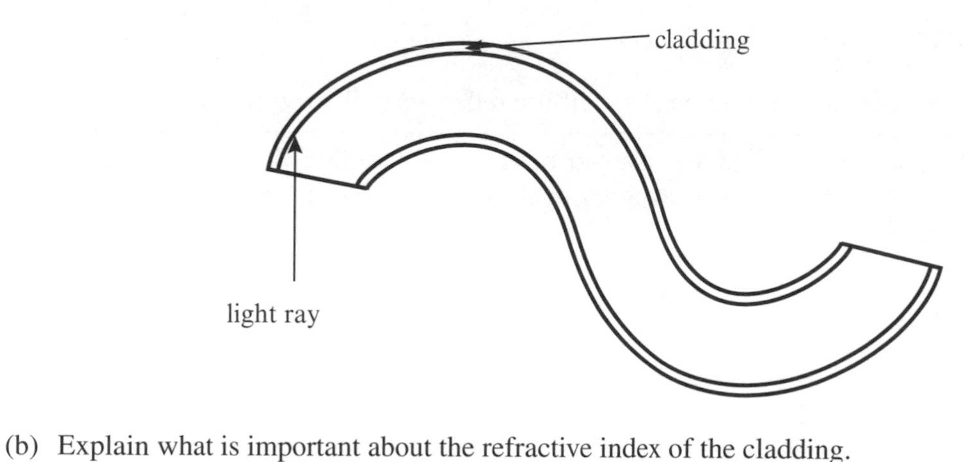

(b) Explain what is important about the refractive index of the cladding.

.... **(1 mark)**

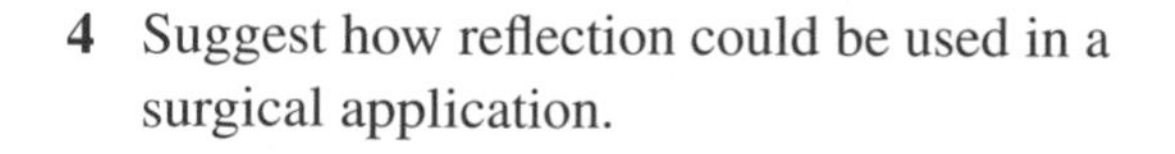

4 Suggest how reflection could be used in a surgical application.

> Consider the equipment a surgeon would use.

....

....

....

....

.... **(4 marks)**

Had a go ☐ Nearly there ☐ Nailed it! ☐

Colour of an object

1 (a) Give three examples of the reflection of waves.

Consider all types of waves, not just light.

1 ....

2 ....

3 .... **(3 marks)**

(b) Give an example of where specular reflection occurs.

.... **(1 mark)**

2 Match the colour of light with the correct description.

Colour
red
blue
yellow
violet

Description
shortest wavelength of visible light
absorbed by a red object
seen through a blue filter
longest wavelength of visible light

(4 marks)

3 Explain why a green object appears green in white light.

Guided

The green object appears green because ....

....

and all other colours .... **(2 marks)**

4 Explain, with the aid of a diagram, how diffuse reflection still obeys the law of reflection.

You will need to draw a diagram of the microscopic, uneven surface of a material that shows how light is still reflected according to θi=θr but from several different angles of the surface.

....

....

....

.... **(5 marks)**

Lenses and power

1 Describe the main difference between the way a converging lens and a diverging lens bend light.

Guided

A converging lens bends the .... **(1 mark)**

A diverging lens bends the .... **(1 mark)**

2 (a) Explain how the focal point of a lens is related to its thickness.

....

.... **(2 marks)**

(b) State what is meant by the power of a lens.

.... **(1 mark)**

3 The diagram below shows two converging lenses.

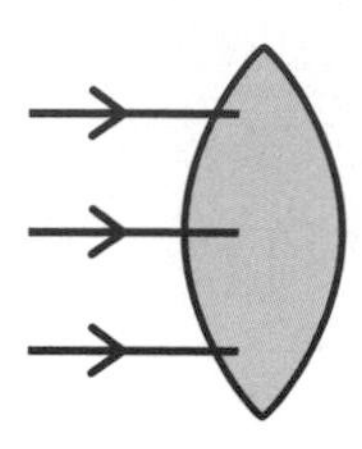

lens 1

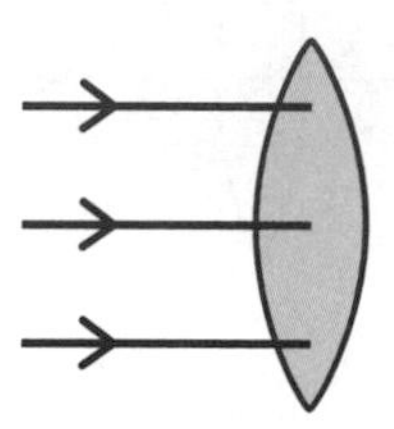

lens 2

(a) Complete the diagrams for both lenses by drawing the rays to converge at a focal point. **(4 marks)**

(b) State which lens is the more powerful.

.... **(1 mark)**

4 The diagram below shows a convex lens and a concave lens.

A convex lens converges the light rays and a concave lens diverges the light rays.

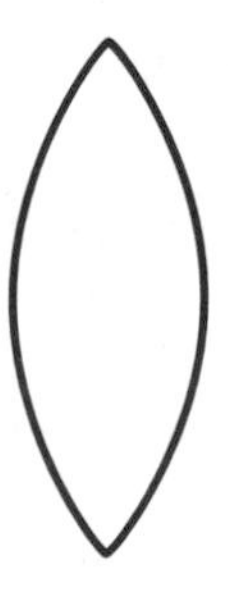

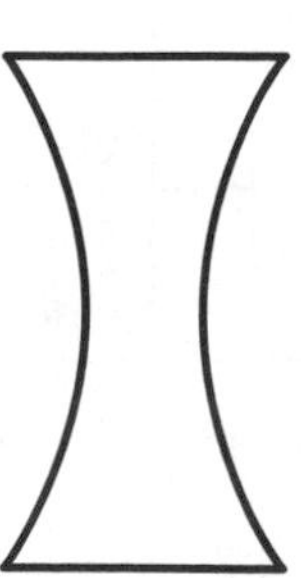

(a) For the convex lens, draw three rays of light to show what the lens does and label the focal point and focal length. **(3 marks)**

(b) For the concave lens, draw three rays of light to show what the lens does and label the focal length. **(3 marks)**

Had a go ☐ Nearly there ☐ Nailed it! ☐

Real and virtual images

1 Explain the difference between a real image and a virtual image. Complete the following paragraph using the words below.

Guided

A real image is an image that can be projected ....

but a virtual image ....

A virtual image is produced when .... **(3 marks)**

2 Give another name for the focal point.

.... **(1 mark)**

3 Add these points to the diagram.

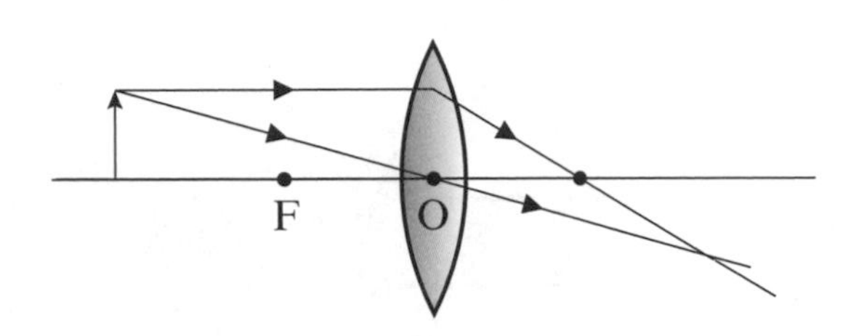

You will need to use your ruler.

(a) 2F on each side of the lens **(1 mark)**

(b) the other focal point **(1 mark)**

(c) an arrow showing the position and type of image

Use the key words in the question to help you find the answers in the table

4 The table below shows conditions for a converging lens.

Object	Image	Real/ virtual	Larger/smaller than object	Upright/ inverted
closer than F	'in front' of lens	virtual	larger	upright
between F and 2F	beyond 2F on the opposite side of the lens	real	larger	inverted
at 2F	at 2F on the opposite side of the lens	real	same size	inverted
further away than 2F	F to 2F on the opposite side of the lens	real	smaller	inverted

(a) State the size of the image relative to the object as the object is placed between F and 2F.

.... **(1 mark)**

(b) State where the object must be for the image type to become a virtual image.

.... **(1 mark)**

(c) Give the position of the object at which the image is real and is the same size as the object.

.... **(1 mark)**

Electromagnetic spectrum

1 Microwaves and ultraviolet are types of radiation. Identify the statement that describes these waves correctly.

☐ **A** Microwaves have a higher frequency than ultraviolet.

☐ **B** Microwaves and ultraviolet are transverse waves.

☐ **C** Microwaves have a shorter wavelength than ultraviolet.

☐ **D** Microwaves and ultraviolet are longitudinal waves. **(1 mark)**

2 Visible and infrared radiation are given out by a candle. Gamma rays are emitted by radioactive elements such as radium.

(a) State two similarities between all waves in the electromagnetic spectrum.

Guided

All parts of the electromagnetic spectrum are .... waves

and they all .... **(2 marks)**

(b) Explain why different parts of the electromagnetic spectrum have different properties.

.... **(1 mark)**

3 The chart below represents the electromagnetic spectrum. Some types of electromagnetic radiation have been labelled.

longest wavelength/
lowest frequency

shortest wavelength/
highest frequency

← radio waves →←C→←infrared→B←ultra-violet rays→←A→←gamma rays→

Name the three parts of the spectrum that have been replaced by letters in the diagram.

A: ....

B: ....

C: .... **(3 marks)**

4 The speed of electromagnetic waves in a vacuum is 300 000 km/s. A radio wave has a wavelength of 240 m. Calculate the frequency of the radio wave.

Remember to convert the units.

frequency = ... Hz **(3 marks)**

Had a go ☐ Nearly there ☐ Nailed it! ☐

Investigating refraction

1 (a) Suggest a method that could be used to investigate the refraction of light using a glass block and a ray box.

....

....

....

....

....

....

.... **(4 marks)**

(b) Explain what conclusion you would expect to find using the method you have outlined in (a). Your answer should refer to the angle of incidence and the angle of refraction.

Guided

When a light ray travels from air into a glass block ....

....

.... **(2 marks)**

(c) (i) Explain what would be observed if the light ray, travelling through the air, entered the glass at an angle of 90° to the surface of the glass.

.... **(1 mark)**

(ii) Explain what would be happening that could not be observed in this experiment.

....

.... **(2 marks)**

2 State three hazards associated with investigating reflection with a ray box and suggest safety measures that could be taken to minimise the risk.

....

....

....

....

.... **(3 marks)**

3 The refraction of light waves through transparent materials can be modelled using a ripple tank.

Describe what changes you would expect to observe in the waves when the depth of the water is made shallower by placing a glass sheet at an angle to the waves.

> Consider the waves being generated in deeper water and then moving into shallower water.

....

....

....

.... **(2 marks)**

Wave behaviour

1 Which of the following is not true about the behaviour of electromagnetic waves?

☐ **A** Electromagnetic waves travel at 300 000 000 m/s in a vacuum.

☐ **B** Electromagnetic waves are transverse waves.

☐ **C** Electromagnetic waves are all transmitted from space through the atmosphere.

☐ **D** Electromagnetic waves change speed in different materials. **(1 mark)**

2 (a) Describe four properties of electromagnetic waves that explain their behaviour.

Recall the properties and nature of waves.

....

....

.... **(4 marks)**

(b) Give examples of two of the properties described in (a).

....

.... **(2 marks)**

3 (a) Explain the differences between microwaves and radio waves that are used for communications.

....

.... **(2 marks)**

(b) How does the wavelength of these waves dictate the way they are transmitted?

Long waves are reflected by the ionosphere (part of the atmosphere), while shorter waves pass through.

Guided

Microwaves sent from the transmitter are ....

.... **(2 marks)**

Radio waves sent from the transmitter are ....

.... **(2 marks)**

4 Complete the following statements about the relationship between radio waves and electrical charges.

As .... charges move up and down a (transmitting) radio aerial, oscillating .... and magnetic fields move from the antenna, across space. When the oscillating electric .... encounters another (receiving) aerial, it causes oscillations in the receiving .... circuits. **(2 marks)**

5 Explain why space-based telescopes have been able to collect extra information about the Universe, beyond that collected by telescopes based on Earth.

....

....

.... **(3 marks)**

Temperature and radiation

1 A cup of coffee is left on the table and cools down. Which statement explains why this happens?

☐ **A** The cup absorbs more radiation than it emits.

☐ **B** The cup emits more radiation than it absorbs.

☐ **C** The cup emits the same amount of radiation as it absorbs.

☐ **D** Radiation is transferred from the surroundings.

(1 mark)

2 A bakery sells freshly baked, warm loaves of bread. One loaf is wrapped in white paper and one loaf is wrapped in black paper. Explain which loaf will stay warm longer.

Guided

The loaf wrapped in ....

because ....

.... **(2 marks)**

3 Explain why a potato baked in the oven at 160 °C will radiate infrared radiation when it is taken out of the oven.

....

....

....

.... **(3 marks)**

4 Astronomers are able to analyse the temperature of stars in the Milky Way by examining the colour and intensity of light that the stars emit. Describe how astronomers can use this data to determine the temperature of a star.

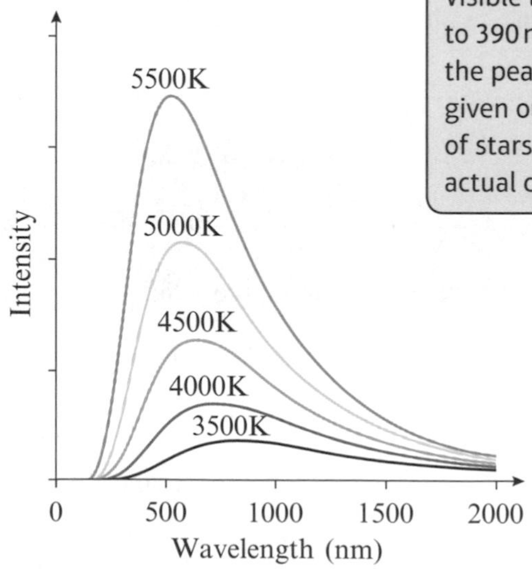

Visible light ranges from 780 nm (red) to 390 nm (violet). The graph shows the peak (main) wavelength of light given out from certain temperatures of stars. You do not need to work out actual colours shown.

Intensity

5500K

5000K

4500K

4000K

3500K

0 500 1000 1500 2000

Wavelength (nm)

....

....

.... **(2 marks)**

Thermal energy and surfaces

1 (a) Describe an experimental method, using the apparatus in the diagram below, to investigate the radiation of thermal energy.

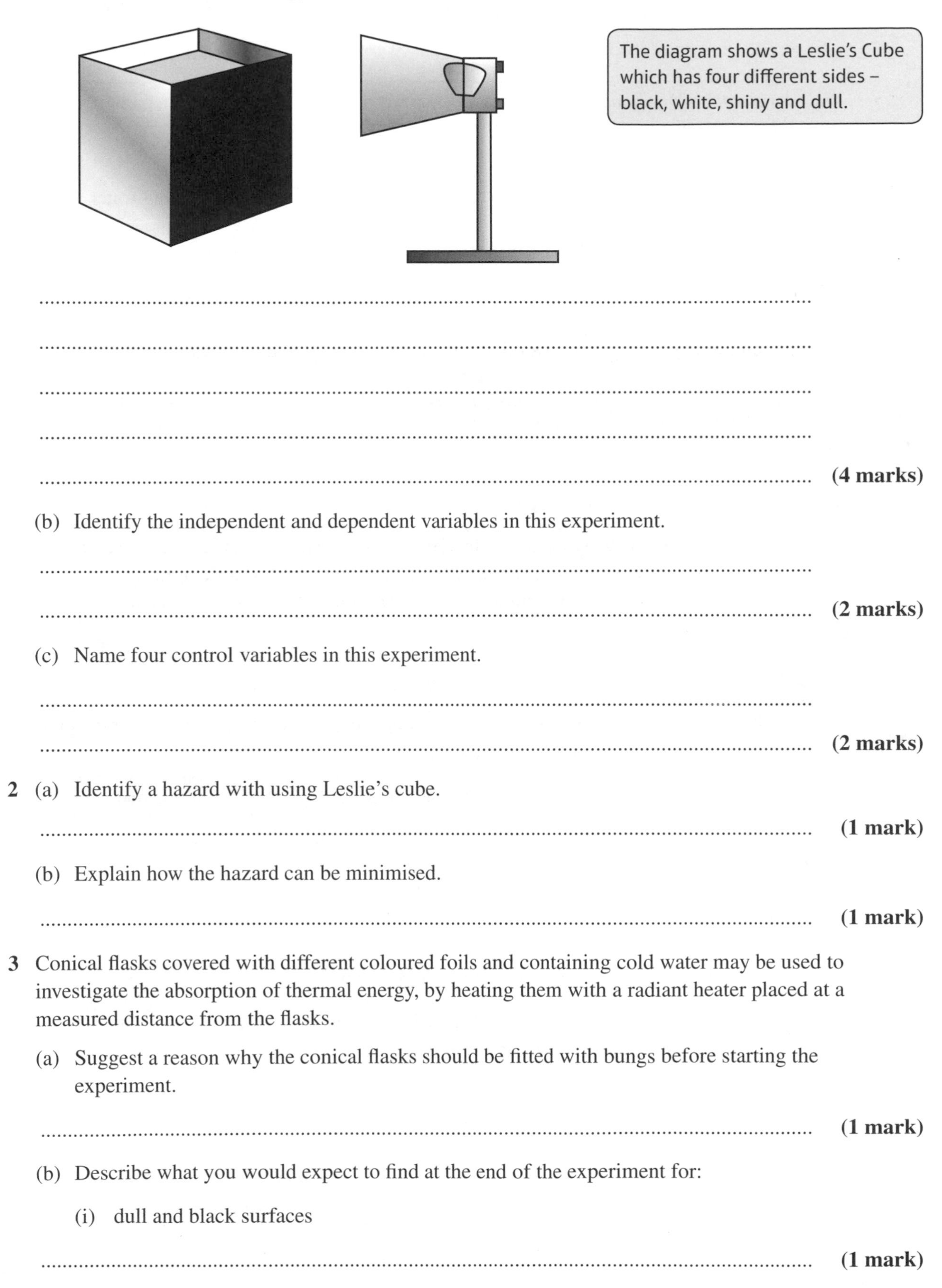

....

....

....

....

.... **(4 marks)**

(b) Identify the independent and dependent variables in this experiment.

....

.... **(2 marks)**

(c) Name four control variables in this experiment.

....

.... **(2 marks)**

2 (a) Identify a hazard with using Leslie's cube.

.... **(1 mark)**

(b) Explain how the hazard can be minimised.

.... **(1 mark)**

3 Conical flasks covered with different coloured foils and containing cold water may be used to investigate the absorption of thermal energy, by heating them with a radiant heater placed at a measured distance from the flasks.

(a) Suggest a reason why the conical flasks should be fitted with bungs before starting the experiment.

.... **(1 mark)**

(b) Describe what you would expect to find at the end of the experiment for:

(i) dull and black surfaces

.... **(1 mark)**

(ii) shiny and light surfaces.

.... **(1 mark)**

Had a go ☐ Nearly there ☐ Nailed it! ☐

Dangers and uses

1 Identify the two correct uses of each type of electromagnetic radiation.

(a) **infrared**

☐ **A** night-vision goggles ☐ **B** broadcasting TV programmes

☐ **C** TV remote control ☐ **D** sun-tan lamps

Remember that you need to choose TWO correct answers for each question part (a), (b) and (c).

(1 mark)

(b) **ultraviolet**

☐ **A** thermal imaging ☐ **B** disinfecting water

☐ **C** cooking food ☐ **D** security marking

(1 mark)

(c) **gamma rays**

☐ **A** sterilising food ☐ **B** communicating with satellites

☐ **C** security systems ☐ **D** treating cancer

(1 mark)

2 Complete the following paragraph by circling the correct word in each case.

Some electromagnetic waves can be dangerous. **Microwaves/Ultraviolet** can **heat/freeze** the water inside our bodies causing significant damage to cells. **Infrared/Visible light** waves transfer **thermal/chemical** energy and can cause burns to skin. **Ultraviolet/Radio** waves can damage **eyes/ears** and can cause skin cancer.

(3 marks)

3 Describe how living cells might be affected following over-exposure to X-rays or gamma rays and what might occur at a cellular level as a result of this.

X-rays and gamma waves move as high-energy ionising photons.

....

....

....

.... **(2 marks)**

4 X-ray scans are often taken when athletes are injured, but X-rays are known to be harmful. Discuss the use of X-rays in the treatment of injuries.

Guided

X-rays are useful because ....

X-rays can be harmful ....

The use of X-rays should be controlled by ....

....

....

.... **(3 marks)**

Changes and radiation

1 Which of the following statements about electrons is true?

☐ **A** Electrons only change orbit when they emit electromagnetic radiation.

☐ **B** Electrons only absorb electromagnetic radiation.

☐ **C** Electrons always change orbit when electromagnetic radiation is absorbed.

☐ **D** Electrons can only orbit the nucleus at defined energy levels within the atom. **(1 mark)**

2 Electromagnetic radiation can be absorbed by electrons that orbit an atomic nucleus.

(a) Explain why different types of electromagnetic radiation have different energies.

....

.... **(2 marks)**

(b) Describe what happens to an electron that absorbs electromagnetic radiation.

Guided

When an electron absorbs electromagnetic radiation ....

.... **(1 mark)**

(c) Describe what happens to an electron that emits electromagnetic radiation.

When an electron emits electromagnetic radiation ....

.... **(1 mark)**

3 The diagram shows different electron energy levels in an atom. With reference to the diagram, explain why only photons of a certain amount of energy (specific wavelength) are emitted when an electron moves to a lower energy level.

$n = 7$
$n = 6$
$n = 5$
$n = 4$
$n = 3$
$n = 2$
$n = 1$

Electrons can only orbit the nucleus of an atom at certain energy levels.

Electrons gain a specific amount of energy from photons to move to a higher energy level.

....

....

....

.... **(3 marks)**

4 Explain why photons can also be emitted from the nucleus of an unstable atom and why these have much higher frequencies than those emitted from orbiting electrons.

....

....

.... **(3 marks)**

Had a go ☐ Nearly there ☐ Nailed it! ☐

Extended response – Light and the electromagnetic spectrum

X-rays and gamma rays are widely used in a number of applications. Compare and contrast these waves and give examples of how they can be used safely in industry.

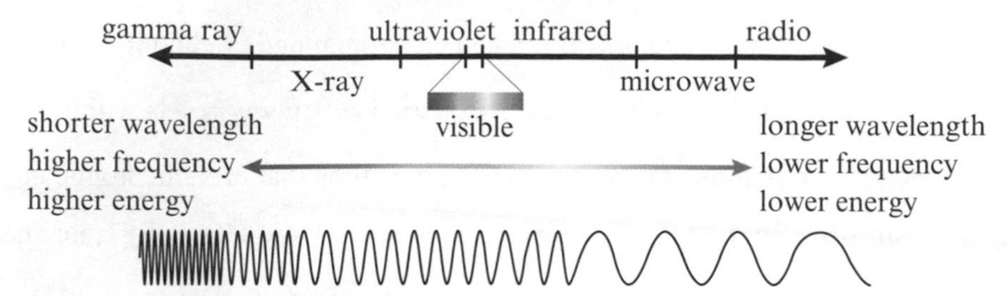

You will be more successful in extended writing questions if you plan your answer before you start writing.

The question asks you to give a detailed explanation of the properties, uses and dangers of X-rays and gamma waves. Think about:

- The types of waves and how you would describe them.
- The dangers of both types of waves and the reasons why they can be dangerous.
- Examples of how the waves are used in medicine.
- Examples of how the waves are used in industry.

You should try to use the information given in the question and in the diagram.

....

....

....

....

....

....

....

....

....

....

....

....

....

....

....

.... **(6 marks)**

Structure of the atom

1 Complete the diagram to show the location and charge of:

(a) protons 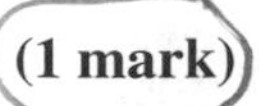 **(1 mark)**

(b) neutrons  **(1 mark)**

(c) electrons.  **(1 mark)**

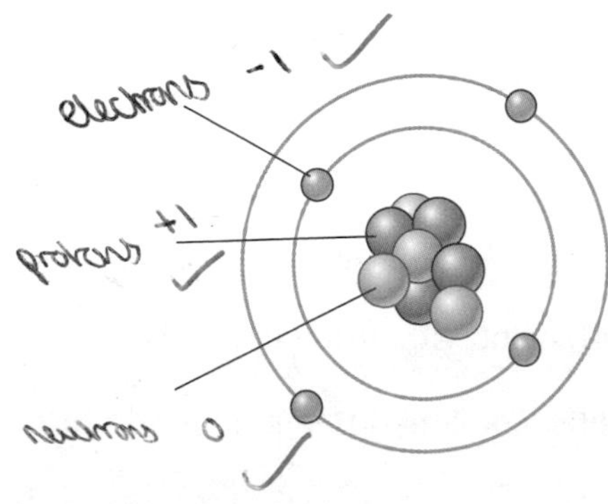

2 (a) Explain why atoms have no overall charge.

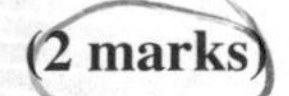 **(2 marks)**

(b) State what will happen to the overall charge if an atom loses an electron.

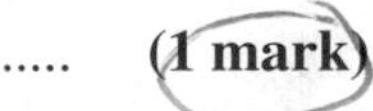 **(1 mark)**

3 (a) State what is meant by the term molecule.

 **(1 mark)**

(b) Give an example of:

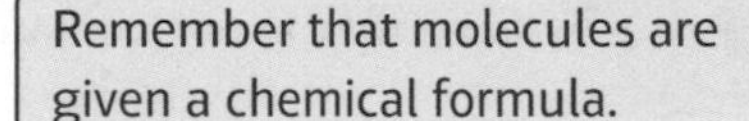

Remember that molecules are given a chemical formula.

(i) a molecule of liquid

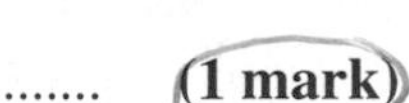 **(1 mark)**

(ii) a molecule of a gaseous element

 **(1 mark)**

(iii) a molecule of a gaseous compound.

(1 mark)

4 The diagram shows an atomic nucleus, an atom and a molecule. Choose the closest approximate size for each and write it under the diagram.

10^{-10} m 10^{-18} m 10^{-9} m 10^{-15} m 10^{-2} m 10^{-6} m

nucleus atom molecule

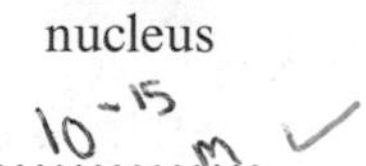

(3 marks)

Had a go ☐ Nearly there ☐ Nailed it! ☐

Atoms and isotopes

1 State what is meant by each term:

Guided

(a) nucleon: the name given to particles in the nucleus ✓ (1 mark)

(b) atomic number: the number of protons (same as the number of electrons) ✓ (1 mark)

(c) mass number: the number of protons and the number of neutrons (the mass of the nucleus) ✓ (1 mark)

2 Identify the correct description of isotopes.

☐ **A** atoms of the same element with different numbers of electrons

☐ **B** atoms of different elements with same numbers of neutrons

☑ **C** atoms of the same element with different numbers of neutrons

☐ **D** atoms of the same element with different numbers of protons ✓

(1 mark)

3 Explain why different isotopes of the same element will still be neutrally charged.

Consider all the particles of the isotopes.

They still have an equal number of protons and electrons. ✓ Neutrons don't have a charge so the number of neutrons doesn't affect the overall charge. ✓ (2 marks)

4 Give the symbol of a potassium atom (K) that has 19 protons and 20 neutrons in its nucleus.

$^{39}_{19}K$ ✓ (2 marks)

5 Identify the correct statement about the relative mass of particles in an atom.

☑ **A** Proton relative mass 1, neutron relative mass negligible, electron relative mass –1. ✗

☐ **B** Proton relative mass 1, neutron relative mass negligible, electron relative mass negligible.

☑ **C** Proton relative mass 1, neutron relative mass 1, electron relative mass negligible.

☐ **D** Proton relative mass 1, neutron relative mass 1, electron relative mass –1.

~~(1 mark)~~

6 The symbols of two isotopes of oxygen are shown below.

$^{16}_{8}O$ $^{18}_{8}O$

Compare the structures of the atoms of the two isotopes.

Consider the particles that are represented by the mass numbers (A) and the atomic numbers (Z).

Both isotopes have 8 electrons so they have 2 shells of electrons, with 2 electrons in the first shell and 6 electrons in the second ~~shell~~. Both isotopes have 8 protons in the nucleus. The first isotope ($^{16}_{8}O$) has 8 neutrons in the nucleus, however the second isotope ~~has~~ ($^{18}_{8}O$) has 10 neutrons in the nucleus. ✓ (3 marks)

Atoms, electrons and ions

1 Select the correct statement about atoms.

☑ **A** Electrons orbit at fixed distances from the nucleus. ✓

☐ **B** Electrons orbit the nucleus at random distances from the nucleus.

☑ **C** An electron can be lost from the atom when it emits electromagnetic radiation.

☐ **D** Electrons move to a lower orbit when they absorb electromagnetic radiation.

(1 mark)

2 Describe what happens to an electron when its atom:

(a) absorbs electromagnetic radiation

Guided

When an atom absorbs electromagnetic radiation the electron gains energy so is promoted to a higher energy level ✓ **(2 marks)**

(b) emits electromagnetic radiation.

When an atom emits electromagnetic radiation the electron loses energy so moves back down to a lower energy level. ✓ **(2 marks)**

3 (a) Write these in the correct part of the table.

F^- Li Na^+ B^+ K^+ Cu

Atoms	Ions
Li Cu	F^- B^+ Na^+ K^+

✓ **(2 marks)**

(b) Explain your choices.

Describe the influence of electrons on both atoms and ions.

Atoms have the same number of protons and electrons so they are neutral and don't have a charge. Ions are atoms that have either lost or gained electrons so they either have more protons than electrons or more electrons than protons and have an overall positive or negative charge. ✓ **(2 marks)**

4 Explain two ways in which a neutral atom can become a positive ion through losing an electron. You may sketch a diagram to help you explain your answer.

Consider the effect of ionising radiation.

An atom can lose an electron by friction
An atom can be made to lose an electron by ionising radiation

(2 marks)

Ionising radiation

1 Select the correct description of an alpha particle.

☐ **A** helium nucleus with charge –2

☑ **B** helium nucleus with charge +2

☐ **C** high-energy neutron

☐ **D** ionising electron

(1 mark)

2 Match the types of radiation with the correct penetrating power.

Type of radiation
alpha
beta minus
neutron
gamma

Penetrating power
low, stopped by thin aluminium
high
very high, stopped by very thick lead
very low, stopped by 10 cm of air

(4 marks)

3 An atom of carbon-14, with 6 protons and 8 neutrons, undergoes beta-minus decay to become an atom of nitrogen, with 7 protons and 7 neutrons.

(a) Give the change in relative atomic mass.

+1 x no change **(1 mark)**

(b) Give another description for a beta-minus particle.

an electron emitted at high speed from the nucleus **(1 mark)**

(c) State the relative ionising property of a beta-minus particle.

low, stopped by thin aluminium **(1 mark)**

4 Identify the type of radiation that would be emitted in each decay.

Remember the law of conservation of mass.

(a) carbon-10 (6 protons, 4 neutrons) → boron-10 (5 protons, 5 neutrons)

beta ~~minus~~ plus **(1 mark)**

(b) uranium-238 (92 protons, 146 neutrons) → thorium-234 (90 protons, 144 neutrons)

alpha **(1 mark)**

(c) helium-5 (2 protons, 3 neutrons) → helium-4 (2 protons, 2 neutrons)

neutron **(1 mark)**

5 Explain why alpha particles have the shortest ionising range in air, compared to other types of ionising radiation.

Guided

Compared to other types of ionising radiation, the chance of collision with air particles ~~is higher in gamma~~ is high

because the alpha particles have a large positive charge. Once an alpha particle has collided with another particle it loses energy. **(3 marks)**

Background radiation

1 The pie chart shows the sources of background radiation. 50% of this comes from the element radon.

(a) Explain what radon is and how it occurs.

(2 marks)

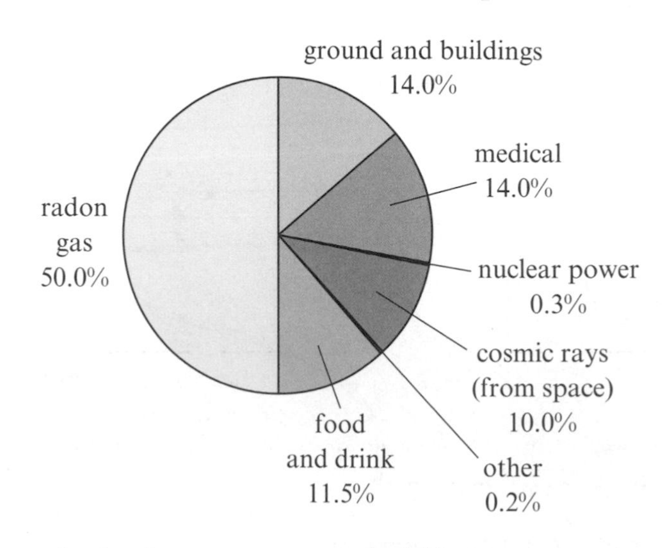

2 Give two reasons why radon levels can vary across the UK.

Guided

Levels can vary because of the different rocks ....

They can also vary .... **(2 marks)**

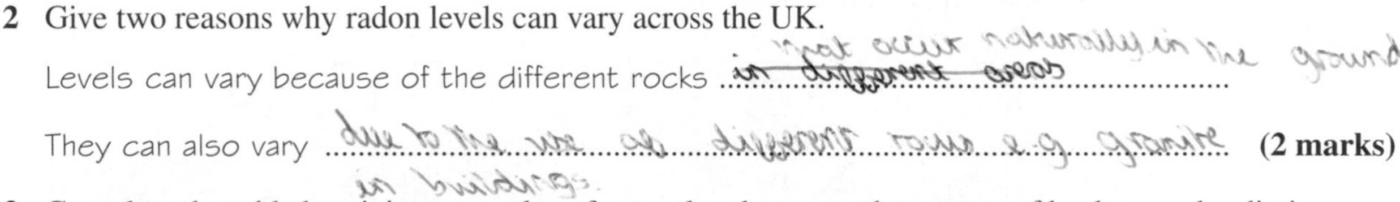

3 Complete the table by giving examples of natural and man-made sources of background radiation.

Sources of background radiation	
Natural	**Man-made**
radon gas	medical
cosmic rays	nuclear power

(2 marks)

4 A scientist in the south-east of England measures the background radiation count three times. Her colleague conducts the same experiment in the south-west. Their results are shown in the table below.

Test number	1	2	3	Average
south-east activity (Bq)	0.30	0.24	0.27	0.27
south-west activity (Bq)	0.31	0.28	0.32	0.30

(a) Calculate the average activity for each sample and write it in the table. **(2 marks)**

(b) State which area has the highest level of background radiation. **(1 mark)**

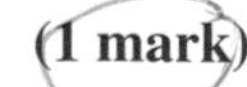

5 (a) Describe how radon gas can get into homes and buildings.

(2 marks)

(b) Explain why inhaling radon can be more dangerous than external exposure.

Consider the type of radiation produced.

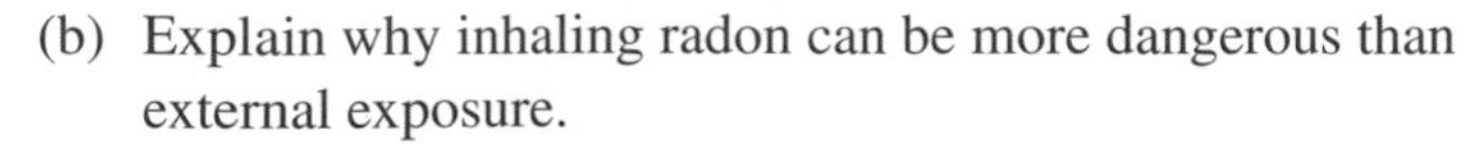

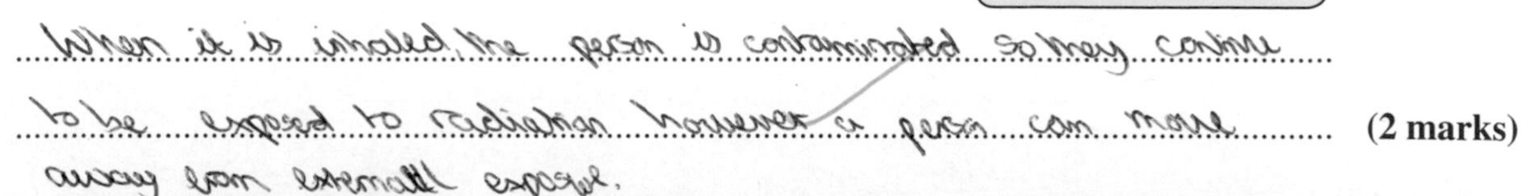

(2 marks)

Had a go ☐ Nearly there ☐ Nailed it! ☐

Measuring radioactivity

1 In 1896 Henri Becquerel discovered that uranium salts led to a darkening of photographic film. Explain how this is used in the nuclear industry today.

Guided

Photographic film is used by nuclear industry workers ....

This monitors levels of ....

.... **(3 marks)**

2 Complete the flow chart below to show how a Geiger–Muller tube detects nuclear radiation.

Guided

A thin wire	→	Atoms of argon are	→	Electrons travel	→	The amount of

(4 marks)

3 A student claims that the more ionising the radiation, the more effective the G–M tube is at detecting levels of radiation. Explain whether the student is correct.

....

....

....

.... **(3 marks)**

4 Explain why photographic film badges have aluminium and lead sheets inserted in front of different parts of the photographic film.

Aluminium will absorb some beta particles and lead will absorb some gamma waves. Absorption will depend on the thickness of the materials.

....

....

.... **(3 marks)**

Models of the atom

1 Compare and contrast the plum pudding and Rutherford models of the atom.

Guided

The plum pudding model showed the atom as ....

....

....

while the Rutherford model showed the atom as ....

....

.... **(4 marks)**

2 Describe the evidence that enabled Rutherford to make his claim about the nucleus.

....

....

....

.... **(3 marks)**

3 (a) What particle did Niels Bohr's model of the atom specifically develop new ideas for?

☐ **A** the electron ☐ **C** the neutron

☐ **B** the proton ☐ **D** the nucleus **(1 mark)**

(b) Explain how the Bohr model of the atom improved on Rutherford's model.

....

....

....

....

....

.... **(4 marks)**

4 (a) Identify the cause of orbiting electrons moving to a higher energy level within the atom.

Electrons orbit only at certain energy levels. They can gain specific amounts of energy from EM radiation.

....

....

.... **(2 marks)**

(b) When an electron loses energy, it can fall to a lower energy level. State the other outcome when an electron loses energy.

....

.... **(1 mark)**

Had a go ☐ Nearly there ☐ Nailed it! ☐

Beta decay

1 Draw lines to link the boxes to complete the sentences about beta-minus and beta-plus decay. One has been done for you.

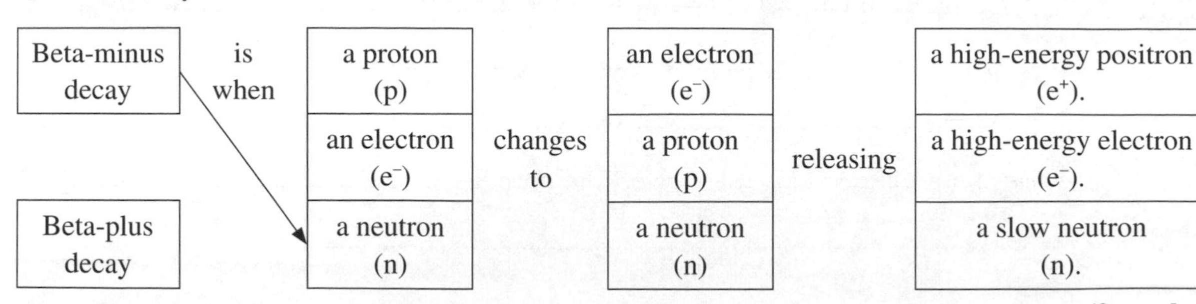

(2 marks)

2 Using the data in the table, complete the equations below.

7 Li lithium 3	9 Be beryllium 4	11 B boron 5	12 C carbon 6	14 N nitrogen 7	16 O oxygen 8	19 F fluorine 9	20 Ne neon 10
23 Na sodium 11	24 Mg magnesium 12	27 Al aluminium 13	28 Si silicon 14	29 P phosphorus 15	31 S sulfur 15	35.5 Cl chlorine 17	40 Ar argon 18

(a) $^{14}_{6}C \rightarrow ^{14}_{\_\_}N + ^{0}_{-1}e$ **(1 mark)**

(b) $^{23}_{\_\_}Mg \rightarrow ^{23}_{11}Na + ^{0}_{1}e$ **(1 mark)**

3 Describe what happens in:

In beta decay neutrons and protons undergo changes producing high energy beta particles. Describe these changes and the particles emitted.

(a) beta-minus decay

....

.... **(2 marks)**

(b) beta-plus decay.

....

.... **(2 marks)**

4 Elements that undergo beta decay are used in archaeological and medical applications. Describe how these professions use beta decay.

Guided

In archaeology beta decay is used to ....

In medicine beta decay is used for .... **(2 marks)**

Radioactive decay

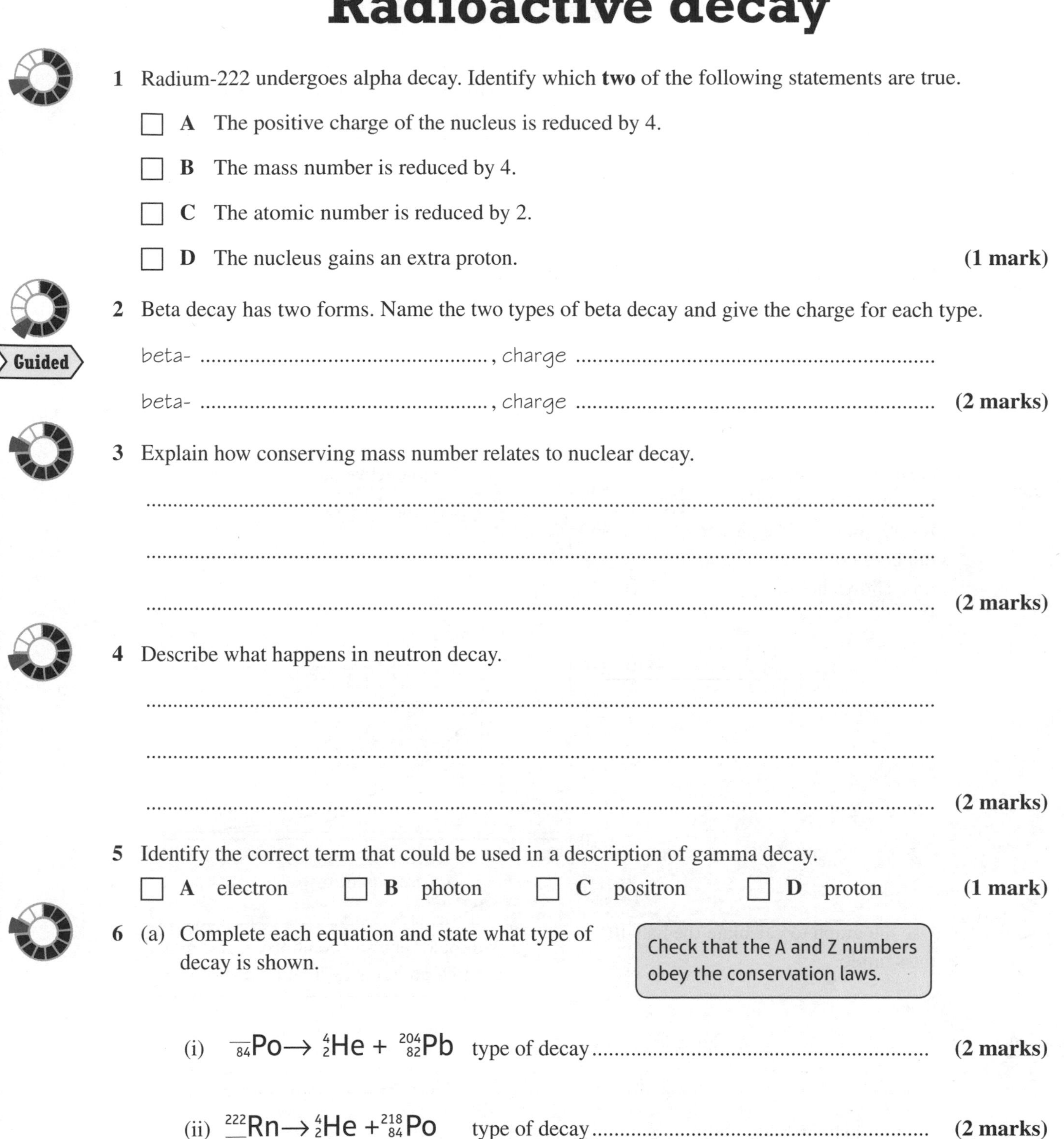

1 Radium-222 undergoes alpha decay. Identify which **two** of the following statements are true.

☐ **A** The positive charge of the nucleus is reduced by 4.

☐ **B** The mass number is reduced by 4.

☐ **C** The atomic number is reduced by 2.

☐ **D** The nucleus gains an extra proton. **(1 mark)**

2 Beta decay has two forms. Name the two types of beta decay and give the charge for each type.

Guided

beta- ...., charge ....

beta- ...., charge .... **(2 marks)**

3 Explain how conserving mass number relates to nuclear decay.

....

....

.... **(2 marks)**

4 Describe what happens in neutron decay.

....

....

.... **(2 marks)**

5 Identify the correct term that could be used in a description of gamma decay.

☐ **A** electron ☐ **B** photon ☐ **C** positron ☐ **D** proton **(1 mark)**

6 (a) Complete each equation and state what type of decay is shown.

Check that the A and Z numbers obey the conservation laws.

(i) ${}^{\_\_}_{84}\text{Po} \rightarrow {}^{4}_{2}\text{He} + {}^{204}_{82}\text{Pb}$ type of decay .... **(2 marks)**

(ii) ${}^{222}_{\_\_}\text{Rn} \rightarrow {}^{4}_{2}\text{He} + {}^{218}_{84}\text{Po}$ type of decay .... **(2 marks)**

(iii) ${}^{42}_{19}\text{K} \rightarrow {}^{0}_{?1}\text{e} + {}^{\_\_}_{20}\text{Ca}$ type of decay .... **(2 marks)**

(iv) ${}^{\_\_}_{4}\text{Be} \rightarrow {}^{1}_{0}\text{n} + {}^{8}_{4}\text{Be}$ type of decay .... **(2 marks)**

(b) Nuclear decay results in a loss of energy from the nucleus. State the reason for this and discuss the energy transfer involved.

....

.... **(2 marks)**

Half-life

1 A sample of thallium-208 contains 16 million atoms. Thallium-208 has a half-life of 3.1 minutes.

Half-life is the time for half the nuclei in a sample to decay, not the time for one atom to decay.

(a) State the number of nuclei that will have decayed in 3.1 minutes.

number of atoms = ... **(1 mark)**

(b) Calculate the number of unstable thallium nuclei left after 9.3 minutes.

number of unstable thallium nuclei left = ... **(2 marks)**

2 A student measured the activity of a radioactive sample for 30 minutes. She plotted the graph of activity against time shown below.

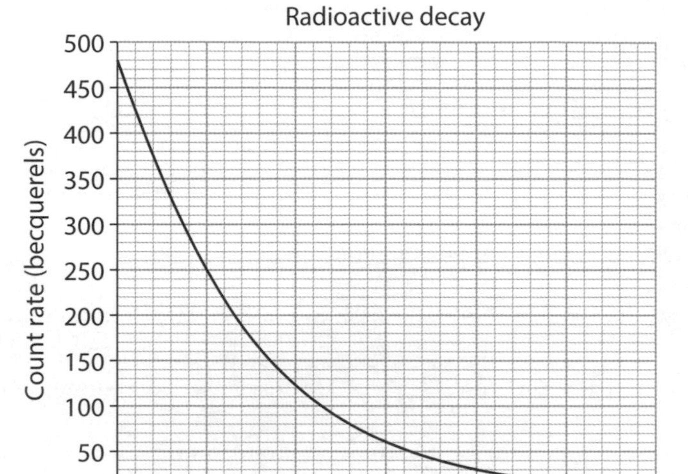

Use the graph to calculate the half-life of the sample.

You could take any point on the line as a starting point for calculating half-life.

Guided

The activity at .... Bq is .... min.

Half this activity is .... Bq, which is at .... min

so the half-life is ....

half-life = ... min **(3 marks)**

3 After the Chernobyl nuclear power station exploded in 1986 a radioactive isotope, caesium-137, fell on northern England and Wales. At one place the activity of a soil sample was 64 Bq in 1986. The radioactivity due to caesium-137 was expected to fall to 32 Bq in 30 years. The level of radioactivity in the soil is higher than predicted. Discuss the factors that may have influenced the accuracy of this prediction.

....

....

.... **(3 marks)**

Uses of radiation

1 Machinery that produces standard sheets of paper uses radiation to check the thickness.

> Recall the properties of alpha, beta and gamma radiation.

(a) Explain which type of radiation is used.

Guided

.... radiation is used because.. **(2 marks)**

(b) A radiation detector measures a sudden drop in the radiation that passes through the paper.

(i) State the most likely cause of this change.

.... **(1 mark)**

(ii) State how the machine responds to the change in activity.

.... **(1 mark)**

2 Gamma rays are used to treat cancers. State the property of gamma rays that makes them useful for treating cancers.

.... **(1 mark)**

3 The diagram of a smoke alarm shows that the radioactive isotope americium-241 is used in the system. It is a source of alpha radiation.

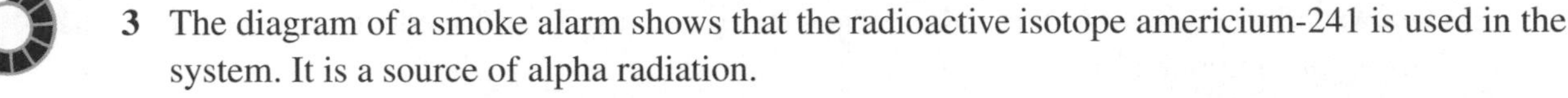

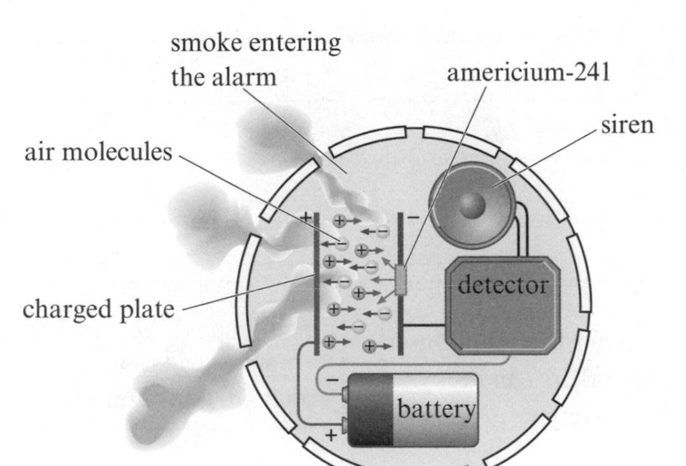

> Alpha particles are absorbed by air and materials such as paper, plastic and metal.

(a) Explain why it is safe to use smoke detectors in the home.

....

.... **(2 marks)**

(b) Explain why the siren sounds when smoke gets into the smoke alarm.

....

.... **(2 marks)**

4 Explain why some surgical instruments are irradiated with gamma rays.

....

.... **(2 marks)**

Dangers of radiation

1 The hazard symbol shown is used to warn that sources of ionising radiation may be present.

Give two places where this symbol may be displayed.

1 ....

2 .... **(2 marks)**

2 (a) State what is meant by the term ionising.

Guided

Ionising means .... **(1 mark)**

(b) Explain why ions are dangerous in the body.

Ions in the body can ....

which can lead to .... **(2 marks)**

3 (a) Describe how employers using radioactive sources can take steps to reduce the exposure of their workers to ionising radiation.

....

....

....

....

.... **(3 marks)**

(b) When risk of exposure to radiation has been minimised through procedure, describe how workers can be monitored to further improve their safety.

Photographic film is an important tool in monitoring levels of exposure to radiation. Think how this is used.

....

....

.... **(2 marks)**

4 Explain why the half-life of a radioactive source is important when considering safety measures.

....

....

.... **(2 marks)**

5 Explain why X-rays are also classed as ionising.

....

....

.... **(2 marks)**

Contamination and irradiation

1 During the First World War (1914–18) soldiers and airmen were issued with watches that had hands and numbers that glowed in the dark. The hands and numbers had been painted with luminous paint that contained radium. Radium was discovered in 1898 and found to be radioactive. In the 1920s many of the women who painted the watches became very ill.

(a) Explain why radium paint was used to paint the watches.

.... **(1 mark)**

(b) Discuss why it was not banned from being used on watches until the 1920s.

Guided Before 1920 the effects of radium ....

....

....

.... **(3 marks)**

2 Draw a line from each term to its correct description.

Term
external contamination
internal contamination
irradiation

Description
A radioactive source is eaten, drunk or inhaled.
A person becomes exposed to an external source of ionising radiation.
Radioactive particles come into contact with skin, hair or clothing.

(2 marks)

3 Give an example of how a person may be subjected to:

(a) external contamination

.... **(1 mark)**

(b) internal contamination.

.... **(1 mark)**

4 Explain why alpha particles are more dangerous from a source of internal contamination than from a source of irradiation.

> Alpha particles can only travel short distances before they collide with another particle and lose their energy. This can have serious consequences near to the body.

....

....

....

....

....

.... **(4 marks)**

Medical uses

1 Identify the type of radiation that is produced by a PET scanner.

☐ **A** alpha particles ☐ **C** positrons

☐ **B** gamma rays ☐ **D** beta particles **(1 mark)**

2 Number these statements in the correct order that describes the production of a PET scan.

A	The radioactive tracer decays to produce positrons.	
B	Gamma rays are produced that are detected by the PET scanner.	
C	A radioactive tracer is injected into the patient.	
D	A computer displays the image on the screen.	
E	When positrons meet electrons they annihilate each other.	

(5 marks)

3 (a) Give the name of the type of chemical used with a PET scanner to help diagnose medical conditions.

....

....

.... **(1 mark)**

(b) State why cyclotrons, used to produce radioactive isotopes for PET scanners, need to be located close to hospitals.

....

....

.... **(1 mark)**

4 Compare and contrast the internal and external approaches to treating cancer tumours.

Consider similarities and differences between the two approaches.

....

....

....

....

....

....

....

.... **(4 marks)**

Nuclear power

1 Give three reasons why nuclear power stations may become a preferred way of generating electricity compared to fossil fuel power stations.

1 ....

2 ....

3 .... **(3 marks)**

2 Describe two types of nuclear reactions that release energy. Give an example for each type.

Guided

Nuclear fusion – ....

This occurs in ....

Nuclear fission – ....

This occurs during ....

....

.... **(4 marks)**

3 Apart from public opinion, identify three practical reasons why building more nuclear power stations may invite opposition from the population.

The half-life of radioactive waste from nuclear power stations can be hundreds, even thousands of years.

....

....

....

....

....

.... **(3 marks)**

4 Explain how it could be argued that global warming could be increased by building more nuclear power stations. Your answer should also explain how the nuclear industry could respond to this issue.

....

....

....

....

....

....

....

.... **(4 marks)**

Nuclear fission

1 Label the particles shown in the diagram of nuclear fission. Use the words below. **(3 marks)**

neutrons	daughter nuclei	uranium-235

uranium-235

energy release

daughter nuclei

neutrons

2 Explain the difference between uranium-235, which can be used in nuclear fission, and uranium-238, which cannot be used.

Uranium-235 is unstable so it splits when a neutron collides however uranium-238 is stable so it doesn't produce ionising radiation.

.... **(2 marks)**

3 Give the function of the following components in the nuclear fission process.

(a) moderator.. **(1 mark)**

(b) control rods.. **(1 mark)**

4 Describe what is meant by the term chain reaction and what is done to control it.

> A chain reaction will happen when increasing numbers of neutrons are released and absorbed by uranium nuclei.

Guided

A chain reaction occurs as neutrons..

....

....

As more than one neutron is released..

....

Control rods are used to .... **(4 marks)**

5 Explain why fast-moving neutrons are not used in the nuclear fission reactors that use uranium-235.

....

....

....

.... **(2 marks)**

Nuclear power stations

1 The diagram shows a nuclear power station.

part: ....
energy store: ....

part: ....
energy store: ....

water

part: ....
energy store: ....

(a) Complete the boxes with the terms below to identify each part of the power station. **(3 marks)**

turbine	reactor	generator

(b) Add these stores of energy to the correct boxes on the diagram:

(i) nuclear store **(1 mark)**

(ii) kinetic store. **(1 mark)**

(c) State the purpose for using nuclear energy in this way.

.... **(1 mark)**

2 Identify the correct particle absorbed by a uranium nucleus during nuclear fission.

☐ **A** fast-moving neutron ☐ **B** plutonium nucleus

☐ **C** high-speed electron ☐ **D** slow-moving neutron **(1 mark)**

3 Explain the process of nuclear fission to release energy from an atom of uranium-235. You may include a diagram to help with your answer.

Guided

Nuclear fission starts with ....

....

....

.... **(4 marks)**

4 Discuss the advantages and disadvantages of using nuclear fission as an energy resource for the generation of electricity.

Nuclear power stations do not release the products of combustion although there are other waste materials to manage.

....

....

....

.... **(4 marks)**

Nuclear fusion

1 Identify the correct process illustrating nuclear fusion.

☐ **A** helium nuclei → hydrogen nuclei

☐ **B** hydrogen nuclei → helium nuclei

☐ **C** uranium nuclei → thorium nuclei

☐ **D** plutonium → uranium nuclei **(1 mark)**

2 In 1989 Martin Fleischmann and Stanley Pons announced that they had evidence that deuterium and tritium nuclei could fuse at a temperature of 50 °C to form helium nuclei. The announcement was made to newspaper reporters who called the idea 'cold fusion'. Since then, other scientists have been unable to repeat this experiment.

Recall the scientific processes in practical work.

(a) Explain why scientists have found it difficult to accept the claims made by Fleischmann and Pons.

Guided

Other scientists have been unable to ....

....

.... **(2 marks)**

(b) Explain why it would be difficult to generate electricity economically using fusion at present.

....

....

.... **(2 marks)**

3 (a) State which extra-terrestrial bodies use fusion processes to sustain their existence.

.... **(1 mark)**

(b) Give two conditions that are necessary for fusion to occur.

....

....

.... **(2 marks)**

(c) With reference to your answer for (b) state what needs to be overcome to allow nuclei to fuse.

.... **(1 mark)**

4 This equation shows the fusion of deuterium and tritium that occurs in stars.

$^{2}_{1}H + ^{3}_{1}H \rightarrow ^{4}_{2}He + ^{1}_{0}n$ + energy

Explain why energy is given out during this reaction.

Guided

When fusion occurs ....

so .... **(2 marks)**

Extended response – Radioactivity

Nuclear power stations currently use nuclear fission as an energy resource. However, nuclear scientists continue to research the possibilities of using nuclear fusion instead.

Explain the process of nuclear fusion and discuss the issues associated with using this as an energy resource.

You will be more successful in extended writing questions if you plan your answer before you start writing.

The question asks you to give a detailed explanation of the process of nuclear fusion and describe some of the difficulties with using this process in the generation of electricity. Think about:

- The type of nuclei used in fusion.
- The conditions under which nuclear fusion occurs naturally.
- The difficulties in simulating natural conditions for fusion.
- The consequences of using nuclear fuel.

You should try to use the information given in the question.

....

....

....

....

....

....

....

....

....

....

....

....

....

....

....

....

....

....

....

....

.... **(6 marks)**

Had a go ☐ Nearly there ☐ Nailed it! ☐

The Solar System

1 (a) Add the missing names of planets, in order of increasing distance from the Sun, starting with Mercury.

Mercury		Earth		Jupiter			Neptune

(4 marks)

(b) Pluto was discovered in 1930 by Clyde W. Tombaugh and it was classified as a planet.

(i) State what Pluto is classified as today.

.... **(1 mark)**

(ii) Explain why you think it was necessary to reclassify Pluto.

Consider the impact of recently improved observation techniques.

....

....

.... **(2 marks)**

2 Ideas about the Solar System have changed over time. Compare and contrast the difference between Greek astronomer Ptolemy's model and Copernicus' model, proposed over 1500 years later.

....

....

....

....

....

.... **(4 marks)**

3 (a) Name the type of celestial body that does not orbit the Sun in the same way as the others.

☐ **A** planet ☐ **C** comet

☐ **B** asteroid ☐ **D** dwarf planet **(1 mark)**

(b) Give a reason for your choice.

.... **(1 mark)**

4 (a) State what is meant by the term orbit.

Guided

The term orbit means .... **(1 mark)**

(b) Explain the difference between a circular orbit and an elliptical orbit.

In a circular orbit..

but in an elliptical orbit .... **(2 marks)**

(c) State what orbits a planet.

.... **(1 mark)**

Satellites and orbits

1 Identify which of the following is not a natural satellite.

☐ **A** Io (moon of Jupiter) ☐ **C** International Space Station

☐ **B** Pluto ☐ **D** Earth **(1 mark)**

2 Give an example of the purpose for each of the following artificial satellites:

> The orbits of geostationary satellites enable them to maintain a position over a specific area of the Earth. The orbits of low-polar satellites enable them to scan the Earth as it turns on its axis.

(a) geostationary satellite

.... **(1 mark)**

(b) low polar orbit satellite.

.... **(1 mark)**

3 (a) State the factor that results in both orbital speed and gravitational field strength combining to produce a stable orbit for a satellite.

.... **(1 mark)**

(b) Describe what would happen to the satellite if a change occurred in:

(i) orbital speed

....

....

.... **(2 marks)**

(ii) gravitational field strength.

....

....

.... **(2 marks)**

4 (a) Explain why a comet will accelerate when approaching the Sun.

.... **(1 mark)**

(b) Describe the orbit of a comet due to the gravitational field of the Sun.

Guided

As the comet approaches the Sun ....

....

....

....

....

....

.... **(5 marks)**

Had a go ☐ Nearly there ☐ Nailed it! ☐

Theories about the Universe

1 In the 1950s some scientists supported the Big Bang theory and others supported the Steady State theory.

(a) Read the following statements and put a tick in the appropriate boxes to show which statements describe the two theories of the Universe.

Statement	Big Bang theory	Steady State theory
A The Universe started as a burst of energy from a tiny point.	✓	
B The Universe has always existed.		✓
C The Universe is expanding.	✓	✓
D New matter is constantly being formed.		✓

(3 marks)

(b) Give a reason why supporters of both theories accept that the Universe is expanding.

Red shift **(1 mark)**

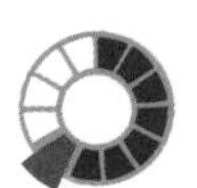

2 Explain why the evidence of cosmic microwave background (CMB) radiation supports the Big Bang theory as the currently-accepted model of the Universe.

Cosmic microwave background radiation (CMBR) originates from radiation created at the beginning of the universe.

As the radio telescopes from all direction show the earth has cooled **(2 marks)**

3 The diagram shows light from the Sun and light from a distant galaxy.

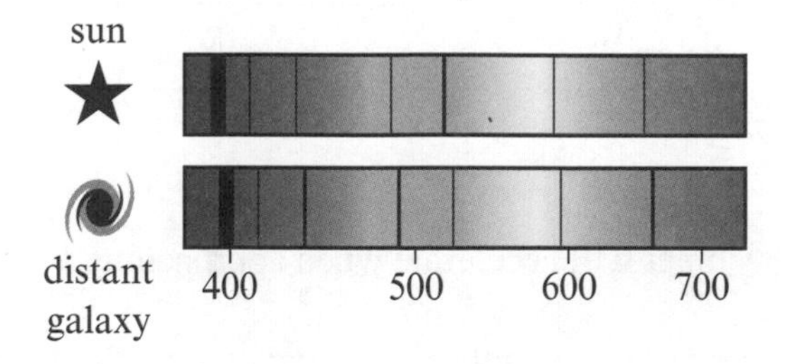

(a) With reference to the diagram, explain the term red shift.

When light from a galaxy is compared with ....

(3 marks)

(b) Describe how the light of a galaxy, observed to have a red-shifted spectrum, gives evidence that the Universe is expanding.

A galaxy with a red-shifted spectrum indicates ....

(3 marks)

Doppler effect and red-shift

1 Identify what happens to light when it is red-shifted.

☐ **A** The light waves become red.

☐ **B** The frequency increases.

☐ **C** It is expanding.

☐ **D** The wavelength increases. **(1 mark)**

2 Explain why the sound of a moving siren is different in pitch than a stationary siren when you listen to it at points A and B.

B

A

Stationary source. The wavelength is the same on both sides.

Moving source. The wavelength is longer behind the source, and shorter in front.

Guided

At point A ....

At point B .... **(2 marks)**

3 An astronomer observes two comets in the Solar System. The light from comet A is red-shifted and the light from Comet B is blue-shifted.

(a) Compare the light from the two comets in terms of wavelength and frequency.

....

.... **(2 marks)**

(b) Describe what the astronomer can tell about the motion of these comets relative to Earth.

....

.... **(2 marks)**

4 Hubble investigated the relationships between the red-shift of light and distant galaxies.

(a) Give the link that Hubble found between observing red-shift from galaxies and how far away they are from Earth.

By observing red-shift, Hubble was able to investigate how fast the galaxies were travelling.

.... **(1 mark)**

(b) Explain the overall conclusion drawn about the nature of the Universe from these observations.

Guided

When the galaxies were plotted on a graph of speed against distance from Earth,

Hubble found that ....

.... **(2 marks)**

Had a go ☐ Nearly there ☐ Nailed it! ☐

Life cycle of stars

1 Name the process that occurs in a star when hydrogen nuclei combine to become helium nuclei.

.... **(1 mark)**

2 Describe what causes a star at the end of the red giant stage to change in size to form a white dwarf.

At the end of the red giant stage the star will still have a huge mass and therefore a gravitational field.

red giant

white dwarf

....

....

....

....

....

....

....

.... **(4 marks)**

3 Identify the true and false statements about the life cycle of stars similar in mass to our Sun by writing T (true) or F (false) in the box.

Statement	T/F
Main sequence stars experience balanced forces.	
White dwarf stars experience nuclear fusion.	
Helium is converted to hydrogen in the nuclear fusion process.	
Stars expand when they lose mass.	
Large amounts of energy are released as helium is produced through fusion.	

(5 marks)

4 Explain why the force that compressed the matter to form a star originally does not cause it to collapse during its life as a main sequence star.

A main sequence star undergoing fusion consequently produces large amounts of thermal energy.

....

....

....

....

....

.... **(3 marks)**

Observing the Universe

1 When exploring the Universe astronomers have used and developed many types of telescope.

(a) State the result of making telescopes that have greater precision.

.... **(1 mark)**

(b) Describe how early astronomers observed the universe.

Guided

Early astronomers used ....

and later .... **(2 marks)**

(c) Give an advantage of using photography when using telescopes.

.... **(1 mark)**

2 Give two examples of why have we learned so much more about the Universe by putting telescopes into space.

Consider how the atmosphere affects what we see.

....

....

.... **(3 marks)**

3 Radio telescopes can be used singly or in an array.
Describe the advantage of using an array of radio telescopes.

An array of telescopes is a collection of telescopes, working together, that can be used as one.

Radio telescopes detect radio waves

....

.... **(1 mark)**

4 Optical observatories are often built in areas of high altitude which creates additional building expenses. Explain why this type of site would be chosen rather than building at sea level.

Optical telescopes detect visible light

....

....

.... **(2 marks)**

5 The Herschel Space Observatory (launched in 2009) orbits the Sun 1.5 million km from Earth. It carries instruments to detect and record infrared radiation with wavelengths between 55 and 625 micrometres (10^{-6} m). These instruments have detected water around a young star. Suggest an explanation why the water could not have been detected by instruments on Earth.

....

....

.... **(2 marks)**

Extended response – Astronomy

Explain what keeps objects in orbit around other objects in the Solar System. Your answer should describe examples of both natural and manufactured bodies that move in orbits.

> You will be more successful in extended writing questions if you plan your answer before you start writing.
>
> The question asks you to give a detailed explanation of the phenomenon of orbiting bodies and give examples of these. Think about:
>
> - The type of orbits that planets and moons exhibit.
> - The balance between orbital speed and gravitational force.
> - The orbits of comets and how they are different from those of planets.
> - Types of artificial satellites and their purpose.
>
> You should try to use the information given in the question.

(6 marks)

Work, energy and power

1 A kettle has a power rating of 2500 W. How much energy does it use in 5 seconds?

☐ **A** 2500 J

☐ **B** 500 J

☐ **C** 500 W

☐ **D** 12 500 J **(1 mark)**

2 Give the energy store that increases in each of these examples:

Energy transfers result in the movement of energy from one store to another.

(a) when a mass is lifted through a height

.... **(1 mark)**

(b) when a pan of water at 20 °C water is heated to 70 °C

.... **(1 mark)**

(c) when an extra cell is added to a circuit.

.... **(1 mark)**

3 A microwave cooker heats a drink in 20 seconds using 15 000 J of electrical energy. Calculate the power of the microwave cooker. State the unit.

Guided

energy transferred = ... J, time taken = ... s

$P = E/t$ = ...

power = ... unit .... **(3 marks)**

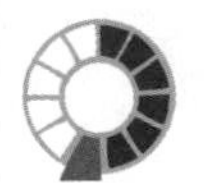

4 A student weighing 600 N climbs 20 stairs to a physics lab. Each stair is 0.08 m high. Calculate the work done by her muscles to climb the stairs. State the unit.

work done = ... unit .... **(3 marks)**

5 A student watches a programme on his television, which has a power rating of 200 W and uses 360 000 J of energy during the viewing. Calculate the time the student spends watching the television.

time taken = ... s **(3 marks)**

Had a go ☐ Nearly there ☐ Nailed it! ☐

Extended response – Energy and forces

Wind turbines are designed to use the kinetic energy of moving air to turn the turbine blades. Explain why this is described as a mechanical process and, therefore, why efficiency is important.

You will be more successful in extended writing questions if you plan your answer before you start writing.

The question asks you to give a detailed explanation of the mechanical processes involved in energy transfers in a wind turbine. Think about:

- The meaning of 'mechanical process'.
- The parts of the process where mechanical processes occur.
- How mechanical processes are inefficient and examples of this.
- Methods of reducing wasted energy due to mechanical processes.

You should try to use the information given in the question.

....

....

....

....

....

....

....

....

....

....

....

....

....

....

....

....

....

....

....

....

.... **(6 marks)**

Interacting forces

1 (a) Give the three types of fields that cause objects to interact with each other **without** making contact.

.... **(3 marks)**

(b) Explain which of these is different from the other two and why.

Two of these fields have opposite poles or charges but one acts in only one direction.

....

....

....

.... **(2 marks)**

2 Which **two** of the following correctly describe the similarities between magnetic and electrostatic fields?

☐ **A** Like poles/charges repel.

☐ **B** Like poles/charges attract.

☐ **C** Opposite poles/charges attract.

☐ **D** Opposite poles/charges repel.

(2 marks)

3 Explain why weight and normal contact force are described as vectors.

Guided

Weight is a vector because ....

....

Normal contact force is a vector because .... **(2 marks)**

....

4 A student pulls along a luggage bag, as shown in the diagram, at constant velocity.

(a) Identify the contact forces for the horizontal motion and state which force is larger.

....

.... **(2 marks)**

(b) Name the balanced contact forces in the vertical direction.

.... **(1 mark)**

Free-body force diagrams

1 Which of the following is **not** one of the steps for resolving a resultant force using a scale drawing?

☐ **A** The angles are drawn with a protractor.

☐ **B** The scale is not important.

☐ **C** The vertical component can be found through measurement.

☐ **D** The horizontal line is drawn to scale. **(1 mark)**

2 The diagram shows a bird of prey of mass 2 kg on a branch of a tree.

Remember to convert the units.

(a) Add arrows to the diagram to show the direction of forces acting on the bird. **(2 marks)**

(b) Add the magnitude and units for each force to the diagram. **(4 marks)**

3 Draw a free-body diagram to represent a cyclist accelerating to the left along a horizontal path. **(4 marks)**

Remember to consider the relative length of the arrows drawn.

4 The diagram shows the horizontal and vertical components of a force.

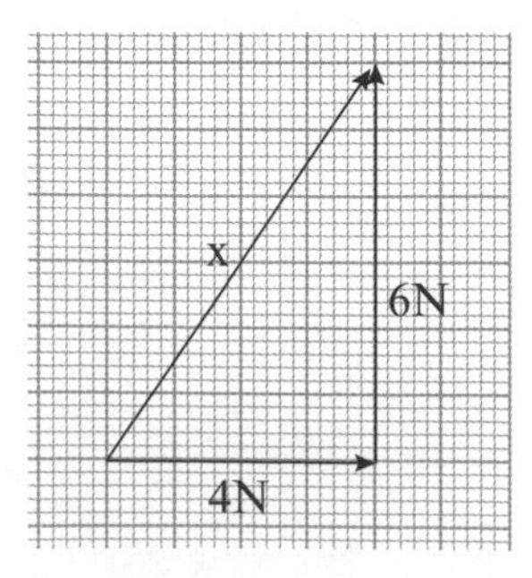

(a) Measure the length of the line of the resultant force x. .... cm **(1 mark)**

(b) Give the magnitude of the resultant force x. .... N **(1 mark)**

Resultant forces

1 Below are diagrams of pairs of forces.

A

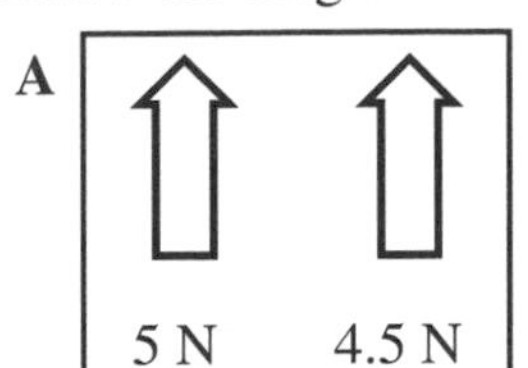

B

6 N 4 N

C

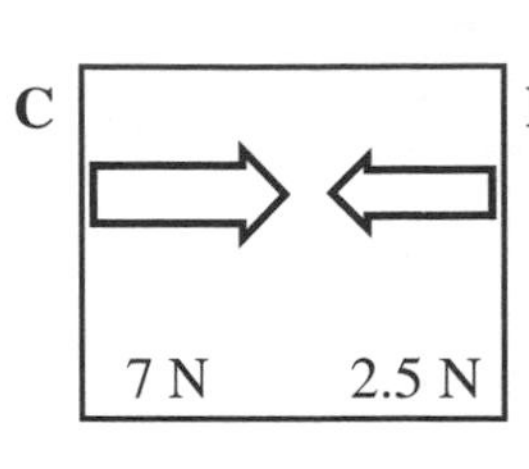

D

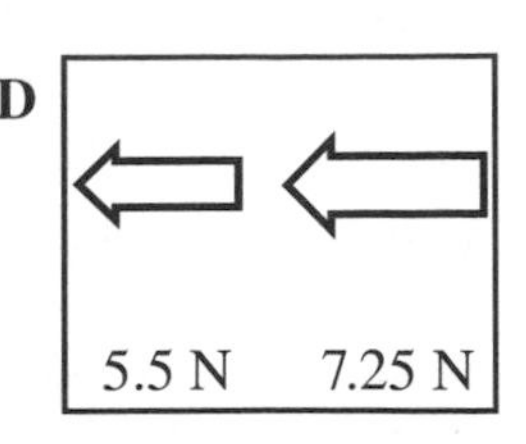

resultant = ... N resultant = ... N resultant = ... N resultant = ... N

direction .... direction .... direction .... direction ....

(a) Calculate the value of the resultant force for each pair. **(4 marks)**

(b) Add an arrow to show the direction of each resultant force. **(4 marks)**

2 The diagram shows two force pairs.

What is the magnitude of the resultant force?

☐ **A** 3 N

☐ **B** 4 N

☐ **C** 5 N

☐ **D** 7 N

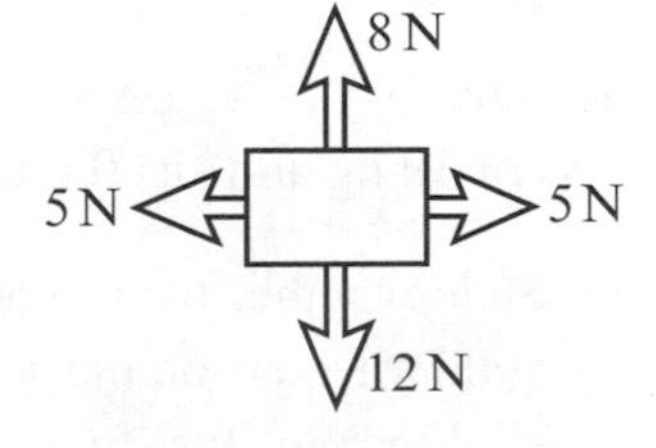

(1 mark)

3 The diagram represents two component forces of a resultant force.

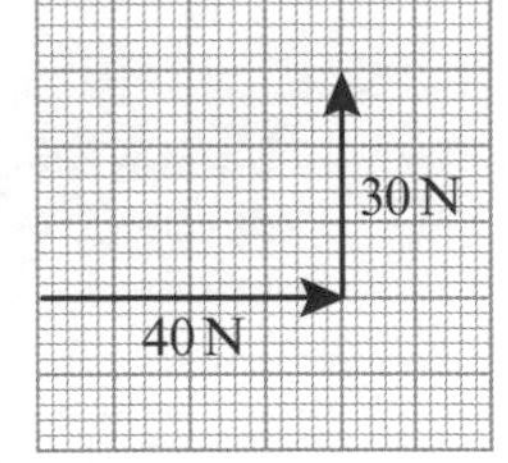

(a) Add an arrow to represent the resultant force.

(1 mark)

(b) Calculate the force represented by the resultant line. .... N **(1 mark)**

4 Determine the resultant force acting on a hockey ball with component forces of 15 N acting horizontally and 6 N acting vertically. Draw a scale diagram to calculate your answer.

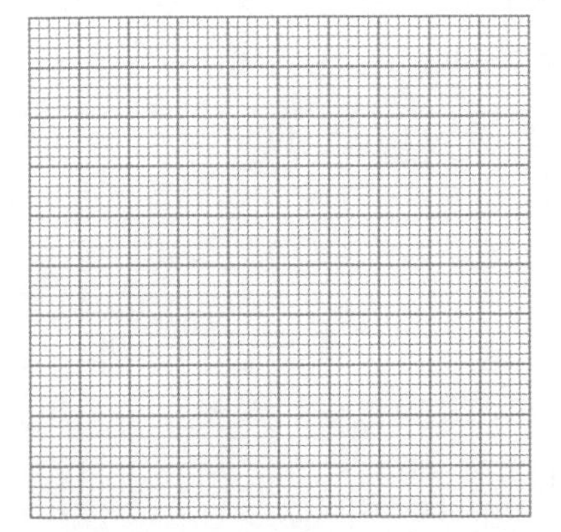

Check your scale before you start to draw the diagram.

(3 marks)

Moments

1 Identify which of the following describes a moment.

☐ **A** a force acting in a clockwise-only direction

☐ **B** a perpendicular force acting at a distance from a pivot

☐ **C** a force acting perpendicular to a surface

☐ **D** a force that causes bending to occur **(1 mark)**

2 Describe the principle of moments for balanced objects.

Guided

When an object is balanced ....

.... **(2 marks)**

3 A screwdriver is used to remove the lid from a tin of paint. Calculate the moment of the screwdriver when a force of 25 N is applied at 0.28 m from the lip of the paint lid. State the unit.

moment = ... unit .... **(3 marks)**

4 Alex and Priya play on a seesaw. Priya has a mass of 25 kg and sits 1.2 m from the pivot. Alex has a mass of 30 kg and sits 0.8 m from the pivot on the opposite side to Priya. Take *g* to be 10 N/kg.

(a) State whether the moments that the children exert on the seesaw are in equilibrium. Justify your answer.

> You will need to calculate the moment for Alex first and then the moment for Priya and compare these. i.e $MA = FA \times dA$ compared with $MP = FP \times dP$.

....

....

.... **(3 marks)**

(b) State how far from the pivot Alex must sit to make the moments balance.

> Rearrange the moment equation to calculate the new distance for Alex i.e. *distance = moment ÷ force*.

distance = ...m **(1 mark)**

(c) Calculate how far Priya must be from the pivot to equal the moment of Alex's original position.

> Rearrange the moment equation to calculate the new distance for Priya i.e. *distance = moment ÷ force* using the first 'moment' you calculated for Alex in (a).

distance = ...m **(2 marks)**

Levers and gears

1 Identify which of the following correctly describes a class 1 lever.

☐ **A** The input and output forces are to the left of the fulcrum.

☐ **B** The input and output forces are either side of the fulcrum.

☐ **C** The input and output forces are to the right of the fulcrum.

☐ **D** A wheel barrow is an example of this. **(1 mark)**

2 Sort these objects into class 1, class 2 and class 3 levers.

broom: fulcrum, input force, output force

scissors: input force, fulcrum

bottle opener: fulcrum, input force, output force

nut cracker: fulcrum, output force, input force

tongs: fulcrum, output force, input force

hammer: input force, output force, fulcrum

Class 1 levers	Class 2 levers	Class 3 levers

(3 marks)

3 Explain the difference between a high gear and a low gear.

For a high gear, the driver gear has ....

than the .... and the ....

For a low gear, the driver gear has ....

than the .... and the .... **(4 marks)**

4 Explain why a cyclist would change from a high gear to a low gear when moving from a horizontal road to a hill.

Consider the force exerted by the cyclist on the pedals.

....

....

....

.... **(2 marks)**

Extended response – Forces and their effects

Balanced and resultant forces are an important consideration when operating a camera drone remotely. To film an aerial sequence, the camera operator flies the drone to the required height for filming and keeps it stationary by controlling the speed of the rotor blades. Explain the balanced or unbalanced forces acting on the camera drone during operation (assume breeze is negligible).

You do **not** need to explain thrust or aerodynamics.

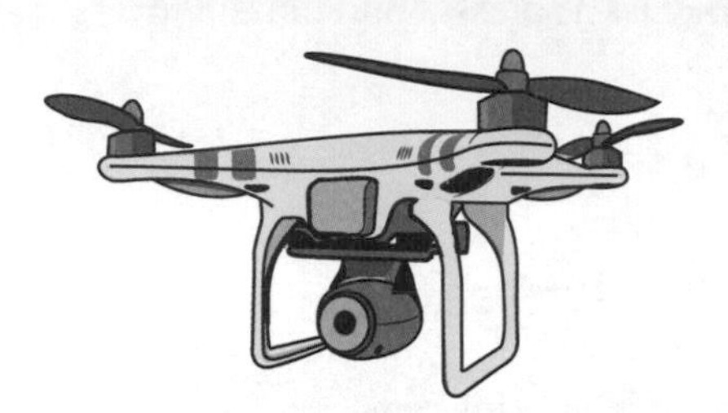

You will be more successful in extended writing questions if you plan your answer before you start writing.

The question asks you to give a detailed explanation of the forces acting on the drone as it is flown to the position for filming. Think about:

- The lift force from the rotor blades.
- How resultant forces lead to motion of an object.
- The vertical and horizontal forces and how they are balanced or unbalanced.
- The forces that the drone must overcome to be able to fly or hover.

You should try to use the information given in the question.

....

....

....

....

....

....

....

....

....

....

....

....

....

....

.... **(6 marks)**

Circuit symbols

1 State the reason why a resistor heats up as a result of an electric current flowing through it.

.... **(1 mark)**

2 (a) Select the **two** components that respond automatically to changes in the environment.

☐ **A** diode ☐ **C** LDR

☐ **B** LED ☐ **D** thermistor **(2 marks)**

(b) Describe how the components you have chosen respond to changes in the environment.

Guided

(i) The .... responds by .... **(1 mark)**

(ii) The .... responds by .... **(1 mark)**

3 Complete the table of circuit symbols below.

Component	Symbol	Purpose
ammeter		
		provides a fixed resistance to the flow of current
		allows the current to be switched on or off

(4 marks)

4 Devise a circuit that could be used to operate a motorised fan to start when the temperature gets too hot. The operator should be able to take readings of the motor voltage and the circuit current. There should also be an option to turn the fan off manually.

Consider whether each component should be connected in series or in parallel.

(6 marks)

Had a go ☐ Nearly there ☐ Nailed it! ☐

Series and parallel circuits

1 (a) Each lamp in these circuits is identical. Write the current for each ammeter on the circuit diagrams.

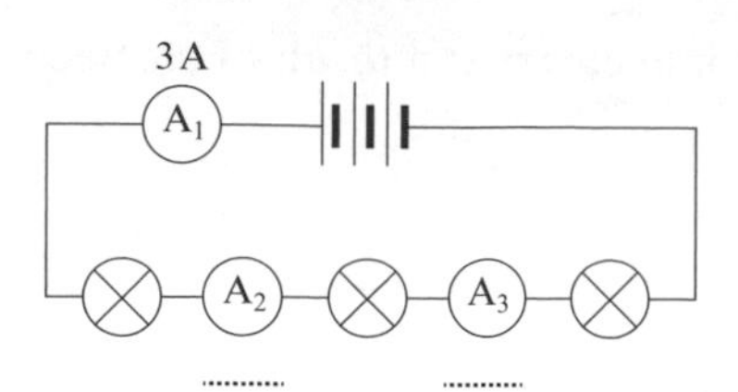

(5 marks)

(b) Explain the rules for current in series and parallel circuits.

Guided

In a series circuit the current ....

In a parallel circuit the current .... **(2 marks)**

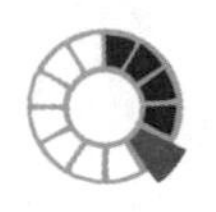

2 (a) Each lamp in these circuits is identical. Write the potential difference for each voltmeter in the circuit diagrams. **(6 marks)**

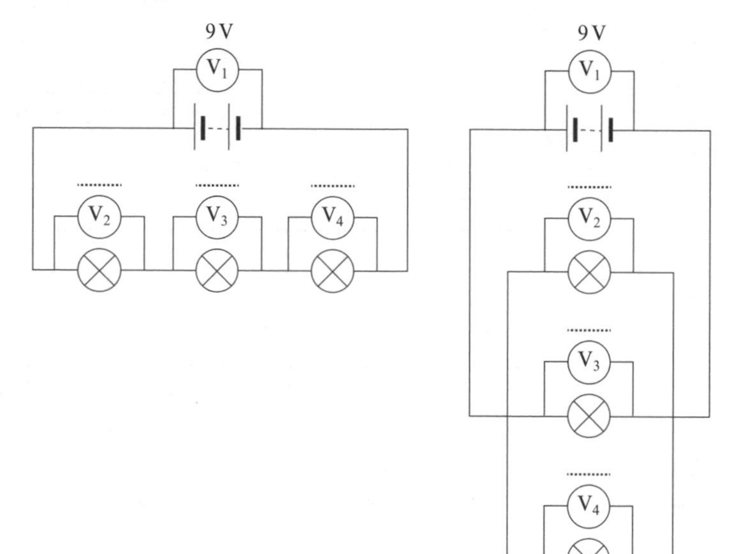

(b) Explain the rules for potential difference in series and parallel circuits.

Guided

In a series circuit the potential difference ....

In a parallel circuit the potential difference .... **(2 marks)**

3 (a) Explain why the electricity supply in a building is connected as a parallel circuit.

Buildings supply electricity for a number of different appliances that are often used at the same time. The supply must be able to support these as well as providing a reliable source of electricity.

....

.... **(2 marks)**

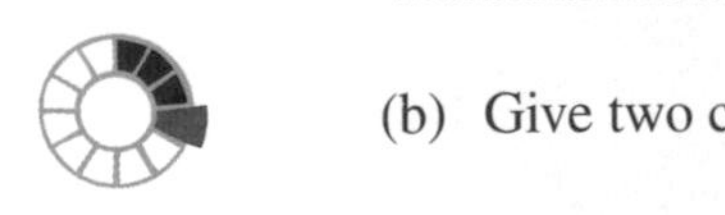

(b) Give two consequences of connecting lamps in a series circuit.

....

.... **(2 marks)**

Current and charge

1 The electric current flowing in a circuit is 4 A.

(a) Explain what is meant by an electric current.

....

.... **(2 marks)**

(b) The current flows for 8 seconds. Calculate how much charge has flowed. Give the unit.

> You may find this equation useful
> $Q = I \times t$

charge = ... unit .... **(3 marks)**

2 The diagram shows a series circuit.

(a) Give the reading on ammeter:

(i) A_1 .... A **(1 mark)**

(ii) A_3 .... A **(1 mark)**

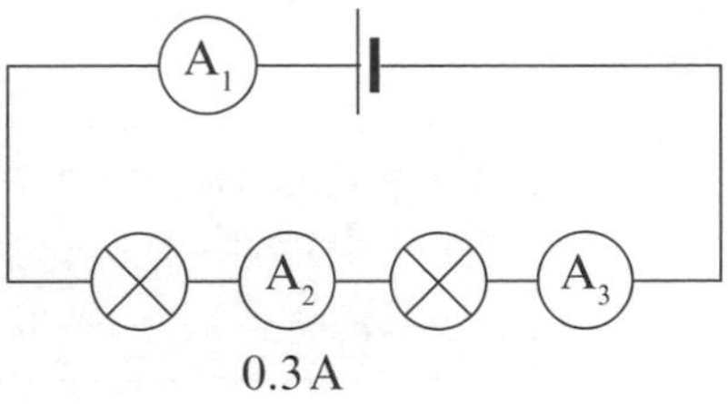

A_1 A_2 A_3 0.3 A

(b) State how you could increase the size of the current flowing through the circuit.

.... **(1 mark)**

(c) Explain why the current measured by ammeter A_2 is the same as A_1 and A_3.

Guided

The electrons move around the ....

so the current leaving the cell is the same as ....

returning to it. **(2 marks)**

3 A student is investigating how current carries charge around the circuit.

(a) Draw a circuit diagram to show how the charge could be measured.

(2 marks)

(b) State what else the student would need to use to collect enough data in order to calculate charge.

.... **(1 mark)**

Had a go ☐ **Nearly there** ☐ **Nailed it!** ☐

Energy and charge

1 State what is meant by current and potential difference. Include the word 'charge' in your answers.

Guided

Current is the ....

Potential difference is the .... **(2 marks)**

2 Calculate the amount of energy transferred to a 9 V lamp when a charge of 30 C is supplied.

Guided

Charge = ... C

Potential difference = ... V

So E = ...

> You may find this equation useful
>
> $E = Q \times V$

energy transferred = ... J **(3 marks)**

3 Calculate the charge needed to transfer 125 J of energy to a string of fairy lights with a total potential difference of 5 V.

> In questions like these, you will need to rearrange familiar equations.
>
> In this case, $E = Q \times V$ will become $Q = E \div V$

charge = ... C **(3 marks)**

4 Calculate how long it takes for 600 J of electrical energy, carried by a current of 0.15 A, to be transferred by a resistor with a potential difference of 20 V.

> This question will need a two-step answer.
> Step one, calculate charge using $Q = E \div V$.
> Step two, calculate the time taken using $Q = I \times t$.
> You will need to be able to rearrange this second equation to give $t = Q \div I$.

time = ... s **(4 marks)**

Ohm's law

1 Which quantity is the ohm (Ω) a unit of?

☐ **A** current ☐ **C** potential difference

☐ **B** energy ☐ **D** resistance **(1 mark)**

2 Explain what Ohm's law means.

Guided

Ohm's Law means that ....

is directly proportional .... **(2 marks)**

3 Calculate the resistance of each resistor.

You may find this equation useful $R = V \div I$

(a) A resistor with a potential difference of 12 V across it and a current of 0.20 A passing through it.

resistance = ... Ω **(2 marks)**

(b) A resistor with a potential difference of 22 V across it and a current of 0.40 A passing through it.

resistance = ... Ω **(2 marks)**

(c) A resistor with a potential difference of 9 V across it and a current of 0.03 A passing through it.

resistance = ... Ω **(2 marks)**

(d) Identify which resistor has the highest resistance.

.... **(1 mark)**

4

Current (A)

0

Potential difference (V)

(a) Sketch two lines on the graph to show two ohmic conductors of different resistances labelled A and B. **(2 marks)**

(b) From your graph, identify which line represents the resistor with the higher resistance.

.... **(1 mark)**

Had a go ☐ Nearly there ☐ Nailed it! ☐

Resistors

1 The diagrams show resistors in series and resistors in parallel. The resistors in both circuits are identical.

Series circuits: $R_T = R_1 + R_2 + R_3$
Parallel circuits: $1/R_T = 1/R_1 + 1/R_2 + 1/R_3$

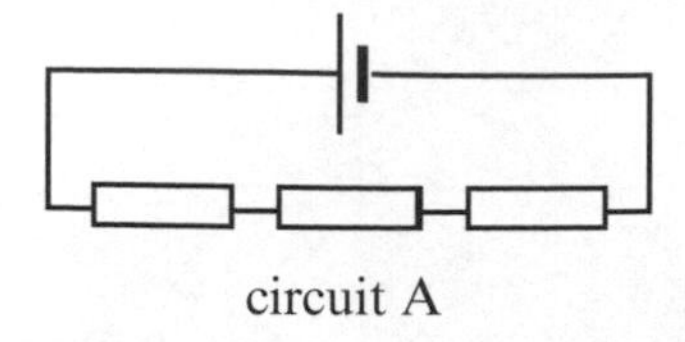

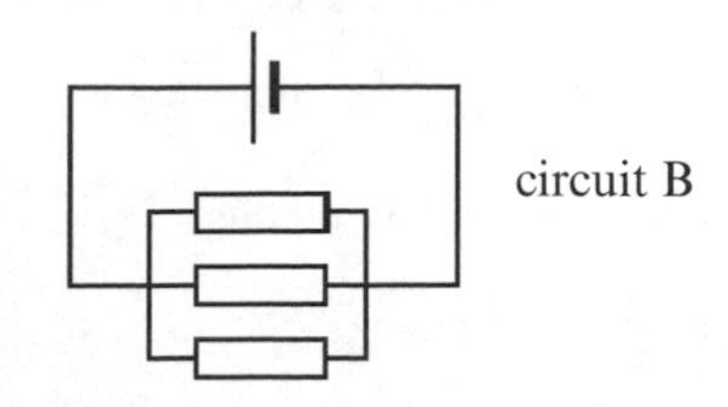

Identify the correct description of the total resistance in circuit B.

☐ **A** It is lower than in circuit A.

☐ **B** It is higher than in circuit A.

☐ **C** It is the same as in circuit A.

☐ **D** The resistance is variable. **(1 mark)**

2 The diagram below shows a 20 Ω resistor, a 30 Ω resistor and a 150 Ω resistor connected in series with identical cells. The current measured by the ammeter is 0.03 A.

(a) Calculate the total resistance of the circuit.

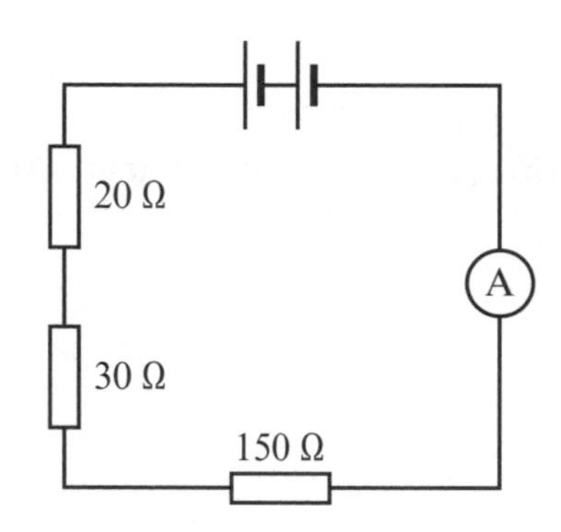

total resistance = ... Ω **(2 marks)**

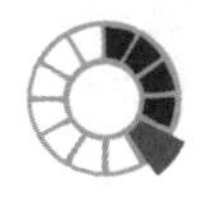

(b) (i) State the rule for potential difference in this type of circuit.

.... **(1 mark)**

(ii) Calculate the potential difference across each cell.

Step 1: Calculate total potential difference of the circuit using $V = I \times R$

where I = ... A and R = (your answer from (a)) .... Ω

So V = ... = ... V

Step 2: Divide the answer at Step 1 by the number of cells in the series circuit.

potential difference = ... V **(3 marks)**

3 The current flowing through the 15 Ω resistor is 0.06 A. The current flowing through the 10 Ω resistor is 0.09 A. Explain why these values for current are different.

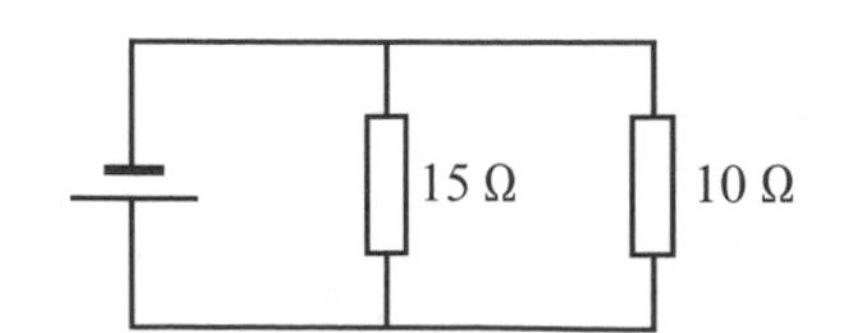

....

.... **(2 marks)**

I–V graphs

1 The graphs below show three types of component.

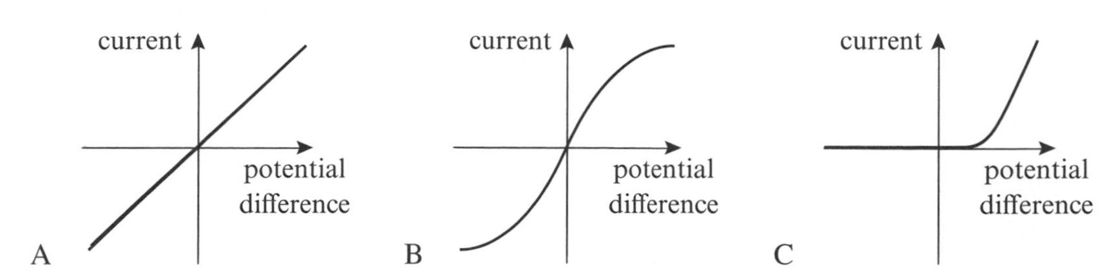

(a) Which graph shows the characteristics of a diode? .... **(1 mark)**

(b) Describe what happens to the current through the component shown in graph A as the potential difference increases.

....

.... **(2 marks)**

(c) Describe what happens to the current through the component shown in graph B as the potential difference increases.

....

.... **(2 marks)**

2 (a) Complete the I–V graphs for a fixed resistor and a filament lamp. **(2 marks)**

Fixed resistor

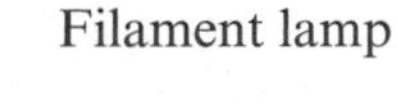

Filament lamp

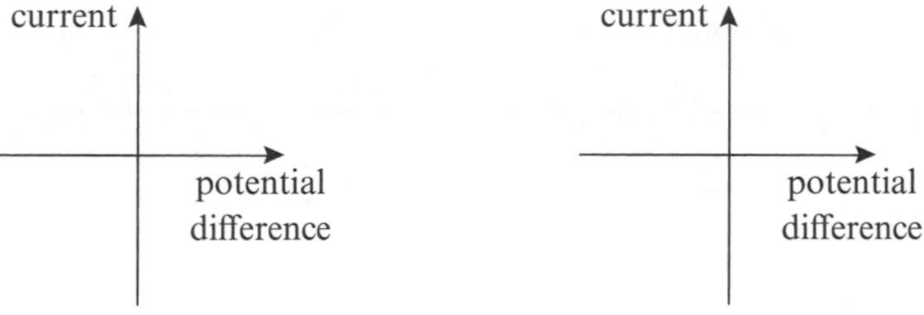

(b) Explain why the filament lamp graph has a different shape to the fixed resistor graph (at constant temperature).

A fixed resistor (at constant temperature) obeys Ohm's Law but a filament lamp does not.

....

.... **(2 marks)**

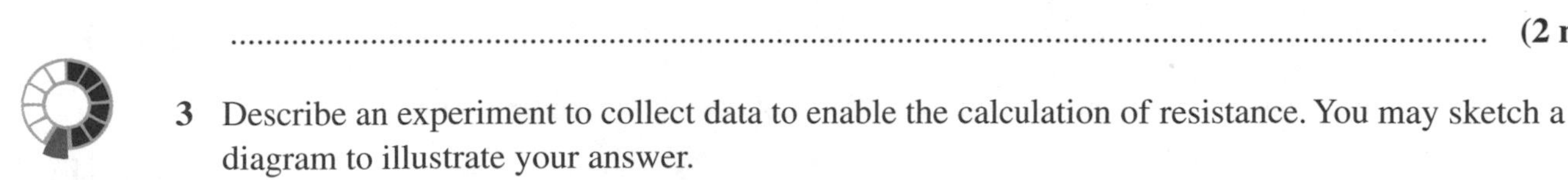

3 Describe an experiment to collect data to enable the calculation of resistance. You may sketch a diagram to illustrate your answer.

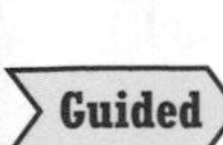

Data can be collected using an ....

and a ....

A ....should be included which will allow

....

Resistance can then be calculated from .... **(5 marks)**

Electrical circuits

1 Electrical circuits can be connected in series or in parallel. Draw a circuit diagram below to show how you could measure the current and potential difference of two resistors connected in a:

Guided

(a) series circuit **(3 marks)**

Remember to include the ammeters and voltmeters in your diagrams.

(b) parallel circuit. **(3 marks)**

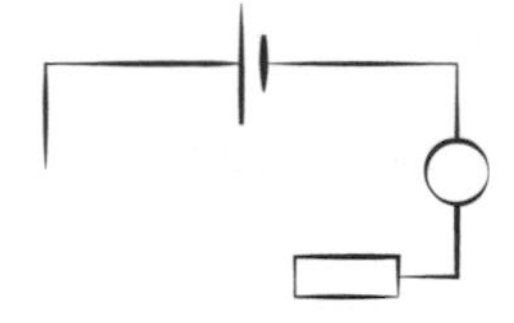

(c) Describe the differences in the current and potential difference in a series and a parallel circuit.

(4 marks)

2 (a) Describe how a circuit may be set up to investigate the relationship between current and potential difference using a filament lamp.

(3 marks)

(b) State the law that may be applied to the data collected in (a) to find resistance.

(1 mark)

(c) Describe how resistance may be found using the data in (a) and a graphical method.

(2 marks)

3 Identify a safety hazard when using resistors.

(1 mark)

The LDR and the thermistor

1 Draw the circuit symbols for the components in the boxes provided.

Light dependent resistor (LDR)	Thermistor

(2 marks)

2 Which variable do you need to change to get a change in the resistance of a thermistor?

☐ **A** current

☐ **B** humidity

☐ **C** light

☐ **D** temperature

(1 mark)

3 Each sketch graph below shows the relationship between two variables.

Recall independent and dependent variables.

(a) Describe how resistance changes with light.

Resistance
Light intensity

Resistance
Temperature

.... **(1 mark)**

(b) Describe how resistance changes with temperature.

.... **(1 mark)**

4 An electric circuit in a car has a lamp connected in series with the battery and a thermistor. The lamp will only light up when the current is above a certain value. Explain the condition necessary for the lamp to light up.

Guided The lamp lights up when the temperature is .... because the current through the lamp and the thermistor ....

.... **(2 marks)**

5 The diagram shows a circuit with a light-dependent resistor. State what happens in terms of resistance and current when the level of light increases.

V_{in} = 12V
R_{top}
1kΩ
LDR
0V
V_{out}
LOW in the light

....

....

....

....

(2 marks)

Had a go ☐ Nearly there ☐ Nailed it! ☐

Current heating effect

1 Which of the following appliances wastes energy due to the heating effect of a current?

☐ **A** lamp ☐ **C** electric fire

☐ **B** kettle ☐ **D** toaster **(1 mark)**

2 Explain how the heating effect occurs in a conductor when a potential difference is applied. Include the words in the box.

lattice	ions	electrons	collisions

Guided

When a conductor is connected to a potential difference ....

....

....

....

....

.... **(4 marks)**

3 Give three examples of how the heating effect of a current can be used in the home.

....

....

.... **(3 marks)**

4 A student claims that it is safe to use his computer, electric fan heater, desk lamp, coffee maker and toaster from the multi-socket extension lead plugged into a single socket, as all the plugs are earthed. Explain what danger the student is risking.

....

....

....

....

.... **(4 marks)**

5 Traditional filament lamps typically only used 5% of their input energy to produce light. Domestic filament lamps were banned by the European Commission, as part of addressing global energy issues, and replaced with more efficient types of lamps such as LEDs. Explain why traditional filament lamps were considered unsuitable for widespread domestic use and the consequence of using lamps such as LEDs.

> You should refer to the form and approximate amount of wasted energy that was transferred by the filament lamps and how low-energy lamps transfer more useful energy.

....

....

.... **(3 marks)**

Energy and power

1 (a) A hotplate is used to heat a saucepan of water. The hotplate uses mains voltage of 230 V. The electric current through the hotplate is 5 A. Calculate the power of the hotplate in watts.

Guided Using the equation for power $P =$ ....

....

power = ... W **(2 marks)**

(b) A mobile phone has a battery that produces a potential difference of 4 V. When making a call it uses a current of 0.2 A. A student makes a call lasting 30 seconds. Calculate the energy transferred by the mobile phone while the call is made. State the unit.

You may find the equation $E = I \times V \times t$ useful.

energy transferred = ... unit .... **(3 marks)**

2 The potential difference across a cell is 6 V. The cell delivers 3 W of power to a filament lamp.

(a) Calculate the current flowing through the lamp.

current = ... A **(3 marks)**

(b) Calculate how much electrical energy is transferred to heat and light in the filament of the lamp when it is switched on for 5 minutes. State the unit.

energy transferred = ... unit .... **(3 marks)**

(c) The lamp in (b) is replaced by a second lamp with a resistance of 240 Ω which draws a current of 0.5 A. Calculate the power rating of the second lamp.

Guided Using the equation $P = I^2 \times R$..

$P =$ ....

power = ... W **(3 marks)**

A.c. and d.c. circuits

1 Circuits can operate using either an alternating current or a direct current.

(a) Explain what is meant by an **alternating** current.

....

.... **(2 marks)**

(b) Explain what is meant by a **direct** current.

....

.... **(2 marks)**

2 Calculate the energy transferred for each of the following appliances:

Remember to convert units where appropriate.

(a) a fan heater (2000 W) running for 15 minutes

energy transferred = ... J **(2 marks)**

(b) a coffee maker (1.5 kW) running for 25 seconds

energy transferred = ... J **(2 marks)**

(c) a tablet charger (10 W) running for 6 hours.

energy transferred = ... J **(2 marks)**

3 A cell in an electric circuit causes charged particles to move along the wires as shown in the diagram.

cell

(a) Describe the current supplied by the cell.

....

....

.... **(2 marks)**

(b) Complete the graph to show what the trace of the current in (a) might look like on an oscilloscope.

Volts

Time

(1 mark)

Mains electricity and the plug

1 (a) Add labels to complete the diagram of a household plug. **(4 marks)**

....
(yellow and green)

....
(brown)

....

....
(blue)

cable grip

outer insulation

(b) Explain which wire the fuse is connected to.

....

.... **(2 marks)**

2 Draw a line from each wire to its correct function.

Wire
brown
blue
green/yellow

Function
Electrical current leaves the appliance at close to 0 V through this wire.
Electrical current enters the appliance at 230 V.
This is a safety feature connected to the metal casing of the appliance.

(3 marks)

3 Explain how a fuse in a plug works.

Guided

When a ....

enters the ....

this produces ....

which ....

The circuit is then .... **(4 marks)**

4 (a) Mains electricity supplies are fitted with a circuit breaker. Explain how a magnetic circuit breaker works.

Magnetic circuit breakers rely on a high current producing a strong magnetic field.

....

....

....

.... **(3 marks)**

(b) Describe how the earth wire in a plug protects the user if the live wire becomes loose.

....

....

....

.... **(3 marks)**

Extended response – Electricity and circuits

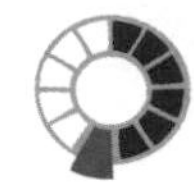

Explain how a circuit can be used to investigate the change in resistance for a thermistor and a light-dependent resistor. Your answer should include a use for each component.

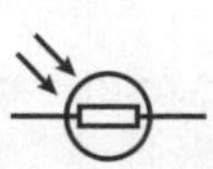

You will be more successful in extended writing questions if you plan your answer before you start writing.

The question asks you to give a detailed explanation of how resistance changes in two types of variable resistor. Think about:

- How resistance in a resistor can be measured and calculated.
- The variable that causes a change in resistance in a thermistor.
- The variable that causes a change in resistance in a light-dependent resistor.
- The consequence to the circuit of a change in resistance in a component.
- Uses for thermistors and light-dependent resistors.

You should try to use the information given in the question.

(6 marks)

Static electricity

1 A student gives a rubber balloon a negative charge by rubbing it on her jumper. She then holds it close to a wall. Explain why the balloon sticks to the wall.

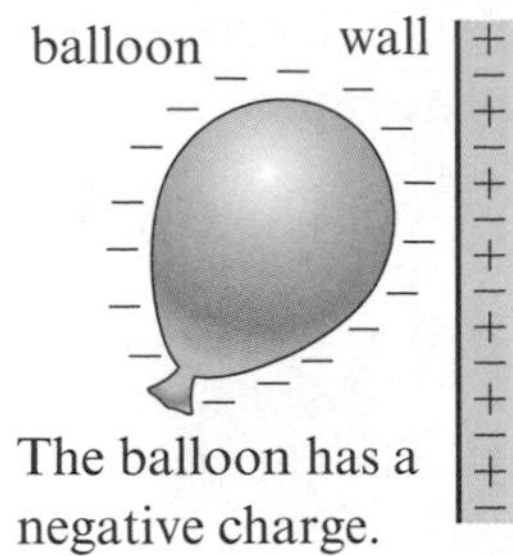

Guided

The student transfers a charge onto the balloon by ....

from .... to the .... The negative charges on the

balloon repel the .... **(4 marks)**

2 Identify the particles that can accumulate to result in a static charge.

☐ **A** electrons on conductors ☐ **C** protons on conductors

☐ **B** electrons on insulators ☐ **D** protons on insulators **(1 mark)**

3 Describe how a material becomes:

(a) positively charged .... **(1 mark)**

(b) negatively charged. .... **(1 mark)**

4 Explain why static charges build up on insulators but not on conductors.

Consider how particles move through the two types of material.

....

....

....

....

....

.... **(4 marks)**

5 In an experiment a student rubs a polythene rod with a cloth.

(a) The student observes that the cloth is attracted to the rod. Explain why this happens.

....

.... **(2 marks)**

(b) The students rubs another rod with the cloth. Describe how the student could show that the two rods have been given the same charge.

....

....

.... **(3 marks)**

Electrostatic phenomena

1 Write the letters of the sentences in the correct order so they describe electrostatic induction by a styrene cup.

A. When the cup is placed near paper an opposite charge is induced.

B. The electrons in the paper are repelled.

C. The cup becomes negatively charged.

D. The positively charged paper is attracted to the negatively charged cup.

E. Rubbing the styrene cup with a cloth transfers electrons from the cloth to the cup.

.... **(4 marks)**

2 A student arrives home and walks down the carpeted corridor to his room, wearing his trainers. When he touches the metal door handle to open the door, he receives a small electric shock. Explain what has happened.

Guided

When the student walked ....

because his trainers ....

When he touched the door handle .... **(3 marks)**

3 Thunderclouds can be very dangerous. They are very large and the thermal convection currents in a thundercloud cause a very large build-up of electrostatic charges.

(a) Explain the processes that lead to lightning occurring.

....

....

....

.... **(4 marks)**

(b) Many tall buildings have a lightning rod made of a thick strip of metal. When a charged thundercloud is nearby, lightning is more likely to strike the rod than unprotected buildings or people. Explain what happens to the lightning rod in an electrical storm where the cloud is negatively charged.

Tall buildings are vulnerable in thunderstorms as they are closer to the negative parts of the thunder clouds than other buildings so low-resistance, conducting 'lightning' rods protect the building.

....

....

....

.... **(3 marks)**

Electrostatics uses and dangers

1 When aircraft are refuelled, a static charge can build up in the fuel pipe.

(a) Describe why this could be dangerous when the aircraft is being refuelled.

Guided

As the aircraft is refuelled ....

and needs to be ....

otherwise the .... **(3 marks)**

(b) Explain how the aircraft is refuelled safely.

....

....

.... **(2 marks)**

2 Static electricity is used in the spray-painting of car bodies and bicycle frames.

(a) Give a reason why the paint and the object being sprayed are given opposite charges.

.... **(1 mark)**

(b) State two advantages of using static electricity for spray-painting.

....

.... **(2 marks)**

3 (a) Explain how electrostatics can be used by farmers to produce healthy crops.

....

....

.... **(2 marks)**

(b) Give two advantages that this method has over other methods of treating crops.

....

.... **(2 marks)**

4 Electrostatic precipitators are used in chimneys to trap smoke particles. Describe how an electrostatic precipitator works.

> Electrostatic precipitators charge smoke particles and collect these before the cleaner gases are released into the atmosphere.

....

....

....

....

.... **(3 marks)**

Electric fields

1 Draw two lines from each point charge to the correct characteristics of their electric fields.

Point charge
positive
negative

Characteristic of electric field
Field acts radially inwards.
A positive charge would move inwards.
A positive charge would move outwards.
Field acts radially outwards.

(4 marks)

2 Explain what is meant by the term electric field.

....

.... **(2 marks)**

3 Draw diagrams to show the electric field of:

(a) a strong positive point charge

(3 marks)

(b) oppositely charged parallel plates close together.

(3 marks)

4 State in which direction (outwards or inwards) the following particles will move when placed in a point source electric field:

(a) (i) a proton in a field generated by a positive point charge .... **(1 mark)**

(ii) an alpha particle in a field generated by a positive point charge .... **(1 mark)**

(iii) a proton in a field generated by a negative point charge. .... **(1 mark)**

(b) Explain why these particles will accelerate in such a field.

When a charged particle is in an electric field it will experience repulsion or attraction depending on its charge.

....

.... **(2 marks)**

5 A student states that 'an electrically charged insulator generates an electric field'. Explain whether the student is right.

Guided

The student is .... because ....

....

.... **(3 marks)**

Extended response – Static electricity

Objects and materials can be given a negative or positive charge. Explain two of the common uses of electrostatic charges.

You will be more successful in extended writing questions if you plan your answer before you start writing.

The question asks you to give two detailed explanations of how static electricity is used in appliances. Think about:

- How electrons can be collected on insulated materials to give a negative charge.
- How electrons can be removed from insulated materials to give a positive charge.
- The resulting effects of repulsion or attraction due to static electricity.
- Uses of static electricity in office appliances.
- Uses of static electricity in industrial processes.

You should try to use the information given in the question.

....

....

....

....

....

....

....

....

....

....

....

....

....

....

....

....

....

....

....

....

....

.... **(6 marks)**

Had a go ☐ Nearly there ☐ Nailed it! ☐

Magnets and magnetic fields

1 Complete the diagrams below to show magnetic field lines for:

(a) a bar magnet **(4 marks)**

(b) a uniform field. **(2 marks)**

bar magnet

N S

uniform field

N S

2 Give two similarities between a bar magnet and the magnetism of the Earth.

Guided

Both a bar magnet and the Earth have ....

They also both have a similar .... **(2 marks)**

3 An electric doorbell uses a temporary magnet to move the hammer, and ring the bell, when the button (switch) is pressed.

soft iron armature
adjusting screw
spring
contact
bell
coil with soft iron core

Explain why a temporary magnet rather than a permanent magnet is used for this application.

....

....

....

.... **(3 marks)**

4 Rajesh carries out an experiment to test for magnetic or non-magnetic materials by moving a permanent magnet near a range of small objects. Some of the objects are attracted to the magnet. Devise a second test that Rajesh could do to further sort the magnetic objects into temporary and permanent magnets using the same permanent magnet.

Think of the difference between a temporary and permanent magnet.

....

....

....

....

.... **(3 marks)**

Current and magnetism

1 The diagram shows a wire passing through a circular card. The **cross** represents the conventional current moving into the page and the **dot** represents the conventional current moving out of the page.

You can use the 'right-hand grip' rule to help you answer this question.

(a)

(b)

(a) Draw lines of flux on each diagram to show the **pattern** of the magnetic field. **(2 marks)**

(b) Draw arrows on each diagram to show the **direction** of the magnetic field. **(2 marks)**

2 The magnetic field around a solenoid is similar in shape to the magnetic field of which of these magnets?

☐ **A** ball magnet

☐ **B** bar magnet

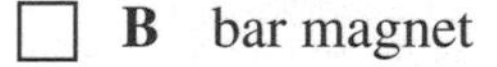

☐ **C** circular magnet

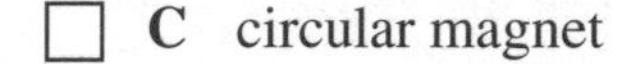

☐ **D** horseshoe magnet **(1 mark)**

3 (a) Give two factors that affect the strength of the magnetic field around a current-carrying wire.

Guided

The strength of the magnetic field depends on ....

and the .... **(2 marks)**

(b) The graphs below show how the strength of a magnetic field varies with two variables. Label the *x*-axes on the diagrams below with either 'current' or 'distance from the wire' to show how they vary with the strength of a magnetic field. **(2 marks)**

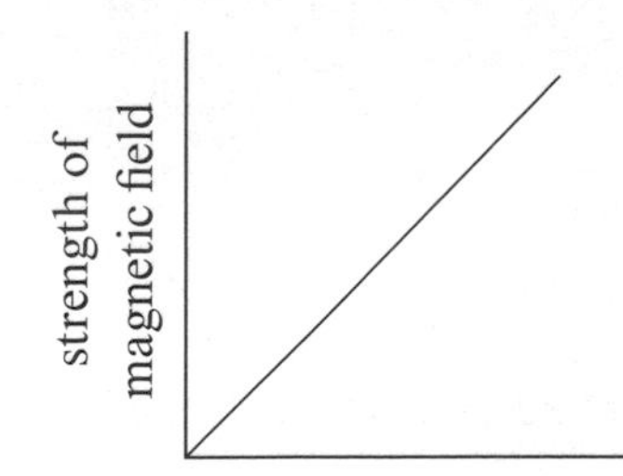

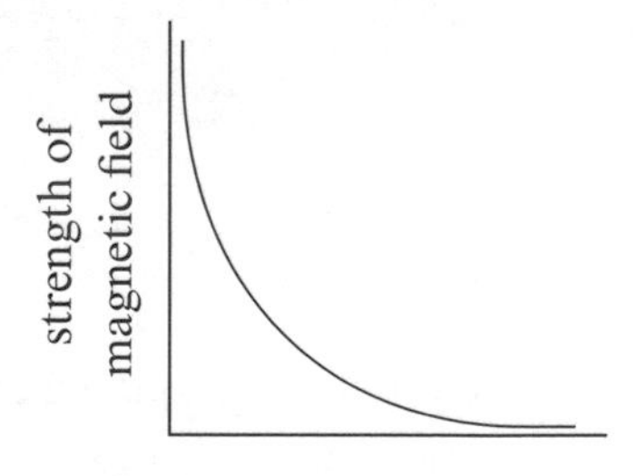

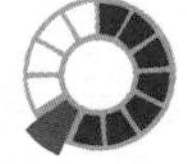

4 In an experiment a student measures the magnetic field strength *B* at a distance of 15 cm from a wire carrying a current of 1.2 A. The experiment is repeated at distances of 7.5 cm and 30 cm.

(a) Give the new value of the magnetic field strength in terms of *B* at:

(i) 7.5 cm from the wire ....

(ii) 30 cm from the wire. .... **(2 marks)**

(b) The student then changes the distance from the wire to 15 cm to take further measurements of the magnetic field strength but changes the current for the first reading to 0.6 A and for the second reading to 2.4 A. Give the new value of the magnetic field strength for:

(i) 0.6 A ....

(ii) 2.4 A. .... **(2 marks)**

Current, magnetism and force

1 A current-carrying wire with a magnetic field is placed in another magnetic field. What is the name of the effect causing the force experienced by the wire?

☐ **A** alternating effect ☐ **C** induction effect

☐ **B** generator effect ☐ **D** motor effect **(1 mark)**

2 Label the diagram showing how Fleming's Left Hand rule may be used to determine the direction of movement of a current-carrying wire in a magnetic field.

....

....

....

(3 marks)

3 Explain how the size of the force in Q2 may be increased.

Guided

The size of the force can be increased by ....

....

or by .... **(2 marks)**

4 Calculate the force acting on a wire of length 30 cm carrying a current of 1.4 A when it is placed in a magnetic field of flux density 0.0005 T. State the unit.

> You may find the equation $F = B \times I \times l$ useful.

force = ... unit .... **(3 marks)**

5 Explain how a motor starts to spin when supplied with an electric current, and how the direction of rotation is maintained.

> A simple motor can be made by placing a coil of current-carrying wire that is able to rotate, between two opposite magnetic poles and adding a commutator.

....

....

....

....

....

....

.... **(5 marks)**

Extended response – Magnetism and the motor effect

Describe how an experiment could show the effect and strength of a magnetic field around a long straight conductor and what would be observed when the circuit was connected.

You will be more successful in extended writing questions if you plan your answer before you start writing.

The question asks you to give a detailed explanation of the magnetic field generated by a current-carrying conductor. Think about:

- How you would safely connect the conductor to enable circuit measurements to be taken.
- The methods you could use to determine the direction of a magnetic field.
- The shape of the magnetic field that you would expect to find.
- How you would interpret the field patterns of current-carrying conductors.
- The variable that would influence the strength of the magnetic field around a current-carrying conductor.
- How the influence of the magnetic field of a current-carrying conductor changes.

You should try to use the information given in the question.

(6 marks)

Had a go ☐ Nearly there ☐ Nailed it! ☐

Electromagnetic induction

1 The diagram shows how a current may be induced in a wire.

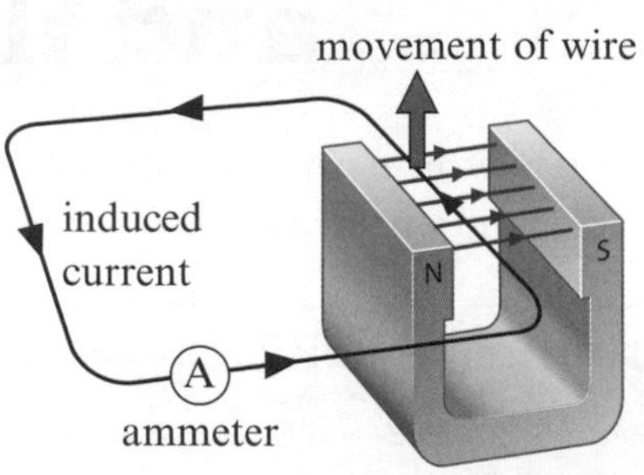

(a) State what must be done to induce a current in the wire that is connected to the ammeter.

.... **(1 mark)**

(b) State how a current in the opposite direction can be induced.

.... **(1 mark)**

(c) State three things that can be done to increase the size of the current.

....

....

.... **(3 marks)**

2 Power stations use large generators (alternators) to generate electricity. Explain why the magnets used in power station generators would need to be very large.

Guided The magnets used in power station generators ....

because they .... **(3 marks)**

3 An electricity generator produces an alternating current.

(a) Sketch a line using the axes below showing how the current changes as the coil of a generator is turned. **(1 mark)**

Current

Time

(b) Explain why this is described as an alternating current. Refer to the graph above.

....

.... **(2 marks)**

Microphones and loudspeakers

1 Identify the characteristic of sound waves that changes the size of the electrical signals produced by a microphone.

☐ **A** amplitude ☐ **C** time period

☐ **B** frequency ☐ **D** wavelength **(1 mark)**

2 Describe the role of the cone in a loudspeaker.

Guided

The cone is moved...

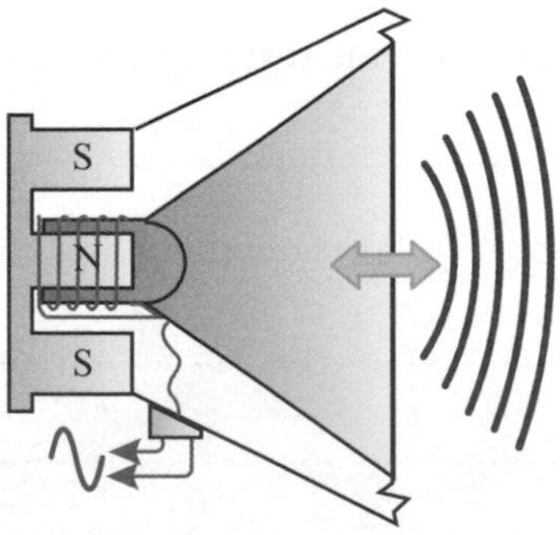

....

....

....

....

....

....

.... **(4 marks)**

3 Describe how a loudspeaker uses magnetic fields to produce a force on the cone.

....

....

....

....

.... **(4 marks)**

4 Explain how the areas of compression and rarefaction in sound waves lead to variation in potential difference in the microphone wire that carries the electrical audio signal.

Use the diagram to help you.

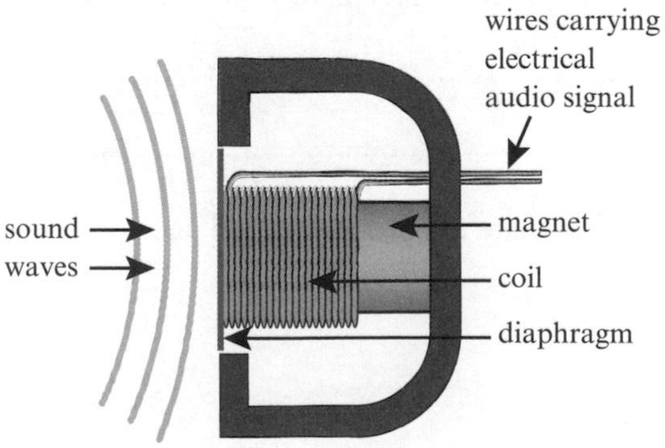

wires carrying electrical audio signal
sound waves
magnet
coil
diaphragm

....

....

....

....

....

....

....

....

.... **(4 marks)**

Transformers

1 Two types of transformers are used in the National Grid. Name them and describe their use.

....

....

.... **(2 marks)**

2 (a) Give the result of increasing voltage before it is transmitted in the National Grid.

.... **(1 mark)**

(b) Explain how this helps to conserve energy.

....

.... **(2 marks)**

3 A laptop computer needs a voltage of 19 V. It is connected to the 230 V mains electricity supply using a transformer with 380 turns on the secondary coil. Calculate the number of turns on the primary coil of the transformer.

Guided

$V_p/V_s = n_p/n_s$, $V_p =$ ...., $V_s =$ ...., $n_s =$ ....

$n_p = n_s \times V_p/V_s =$

Rearrange the equation and substitute the known values.

number of turns on primary coil = ... **(3 marks)**

4 A transformer has 600 turns on its primary coil and 20 turns on its secondary coil. The primary voltage is 360 V.

(a) Explain what type of transformer it is.

....

.... **(2 marks)**

(b) Calculate the secondary voltage.

secondary voltage = ... V **(3 marks)**

Transmitting electricity

1 The National Grid transmits electricity from power stations at 400 000 volts (400 kV).

(a) Explain why this voltage is used to transmit electricity long distances.

Remember that increasing the voltage decreases the current.

....

.... **(3 marks)**

(b) State one hazard of transmitting electricity at 400 000 V.

.... **(1 mark)**

2 Identify the correct link between parts of the National Grid and their functions.

Part of National Grid
step-down transformer
National Grid system
power station
step-up transformer

Function
transmits electrical energy
decreases voltage
increases voltage
generates electrical energy

....

.... **(1 mark)**

3 A power station generates an electric current of 20 000 A at a voltage of 25 kV. Calculate the power generated in kilowatts.

power generated = ... kW **(2 marks)**

4 Describe two ways in which energy losses from the National Grid may be reduced.

....

.... **(2 marks)**

5 Transformers are used at various places in the National Grid. Describe the role of transformers.

Guided

Step-up transformers are used to ....

as it ....

Near homes ....

to reduce ....

.... **(4 marks)**

Extended response – Electromagnetic induction

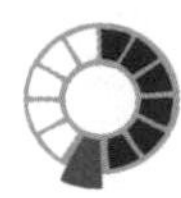

Describe how electromagnetic induction occurs in microphones and headphones.

You will be more successful in extended writing questions if you plan your answer before you start writing.

The question asks you to give two detailed explanations of electromagnetic induction in (a) microphones and (b) headphones. Think about:

- How sound waves are received (microphones).
- The induced electric current (microphones).
- The effects of changing current (microphones).
- The effects of changing current (headphones).
- The induced magnetic field (headphones).
- How sound waves are produced (headphones).

You should try to use the information given in the question.

....

....

....

....

....

....

....

....

....

....

....

....

....

....

.... **(6 marks)**

Changes of state

1 Draw a line from each property to the correct state of matter, and from the state of matter to the correct intermolecular forces. Two lines have been drawn for you.

Property	State	Intermolecular forces
Particles move around each other.	solid	some intermolecular forces
Particles cannot move freely.	liquid	almost no intermolecular forces
Particles move randomly.	gas	strong intermolecular forces

(2 marks)

2 (a) Describe a feature that the three states of matter have in common.

.... **(1 mark)**

(b) Describe two significant differences between the states of matter.

Different states of matter exist due to differences in energy stores which affect the way the particles behave.

....

.... **(2 marks)**

3 Which statement describes the energy change that takes place when ice melts and then refreezes?

☐ **A** Energy is transferred to surroundings → further energy is transferred to surroundings.

☐ **B** Energy is transferred to the ice → energy is transferred to surroundings.

☐ **C** Energy is transferred to surroundings → energy is transferred to the ice.

☐ **D** Energy is transferred to the ice → energy remains in the system. **(1 mark)**

4 Explain why the temperature stops rising when a liquid is heated to its boiling point and heating continues.

Guided

At boiling point the ....

....

....

.... **(3 marks)**

5 When water is put into the freezer and turns to ice at 0 °C, explain what happens, in terms of the energy stored, as the temperature continues to fall to −18 °C.

Remember that the energy stores change as temperature falls.

....

....

....

.... **(3 marks)**

Density

1 A block of wood has a mass of 4000 g and has a volume 5000 cm^3. Calculate its density.

density of wood block = ... g/cm^3 **(3 marks)**

2 Select the **two** correct statements for density.

☐ **A** Density is constant for a material at constant temperature.

☐ **B** Density is related to the atomic packing of a material.

☐ **C** Density changes with increased mass of a material.

☐ **D** Density is calculated using force and volume. **(2 marks)**

3 A metal block measuring 10 cm × 25 cm × 15 cm has a density of 3 g/cm^3. Calculate the mass of the block. Give your answer in kilograms.

Work through the calculation first, then convert mass to kilograms at the end.

mass of metal block = ... kg **(4 marks)**

4 Marco says that all liquids must have lower densities than solids because liquid particles have more kinetic energy and so liquids take up more volume per unit mass than solids. Ella disagrees and says that because solid icebergs float on liquid water, she thinks that Marco must be wrong. Discuss the scientific approaches of these students.

Guided

Marco has approached this problem by ....

....

....

Ella has approached this problem by ....

....

Both students should ....

.... **(5 marks)**

Investigating density

1 When determining the density of a substance you need to measure the volume of the sample.

(a) State which other quantity you need to measure.

.... **(1 mark)**

(b) Give an example of how you could measure this quantity.

.... **(1 mark)**

2 The volume of a solid object may be determined by two methods.

(a) Describe both methods.

....

.... **(2 marks)**

(b) Explain why one method may be preferable over the other.

....

.... **(2 marks)**

3 (a) Describe the method that can be used to find the density of a liquid.

Guided

Place a measuring cylinder on a ....

Add the ....

....

Record the mass of the .... **(3 marks)**

(b) Describe the technique to read the volume of the liquid accurately.

....

.... **(2 marks)**

(c) Calculate the density of a liquid with a mass of 121 g and a volume of 205 cm^3. Give the unit.

You may find this equation useful
density = mass ÷ volume $(\rho = m \div V)$

density = ... unit .... **(2 marks)**

(d) Identify one safety precaution that should be taken when measuring the density of liquids and suggest a method for reducing the risk of harm.

....

....

.... **(1 mark)**

Energy and changes of state

1 State what is meant by the term specific heat capacity.

.... **(1 mark)**

2 Calculate how much energy is required to heat 800 g of water from 30 °C to 80 °C. Take the specific heat capacity of water to be 4200 J/kg °C.

energy required = ... J **(3 marks)**

3 Calculate the amount of energy needed to melt 25 kg ice. Take the specific latent heat of fusion of water to be 336 000 J/kg.

energy required = ... J **(2 marks)**

4 (a) Add the following labels to the graph: melting, boiling. **(1 mark)**

(b) Explain what is happening at these stages to result in no rise in temperature.

Consider the bonds between particles.

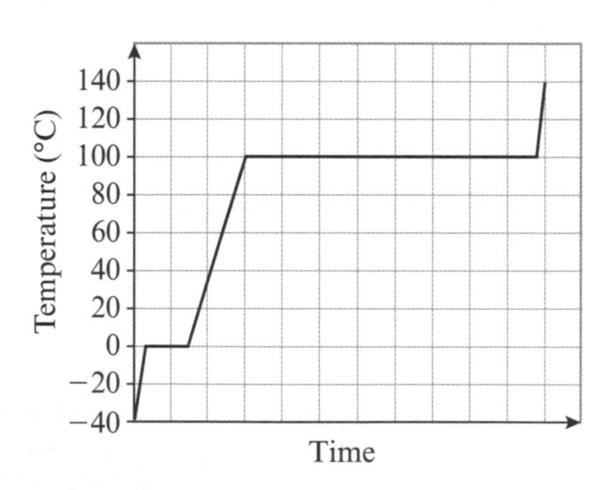

....

.... **(2 marks)**

5 Calculate the specific heat capacity of an 800 g block of metal which is heated for 9 minutes with a current of 2.4 A at 12 V. The temperature increase is 25 °C.

Check the units of the quantities before you substitute into the equation.

specific heat capacity = ... J/kg °C **(4 marks)**

Thermal properties of water

1 Water is widely used in cooling systems because of its relatively high specific heat capacity compared to some other liquids.

(a) State the definition of the term specific heat capacity.

.... **(1 mark)**

(b) Give the equation for specific heat capacity.

.... **(1 mark)**

2 (a) Describe an experiment that could be set up to measure the specific heat capacity of water using an electric water heater, a beaker and a thermometer.

Remember 'pre-experiment' steps e.g. zero the balance to eliminate the mass of apparatus before measuring substances, take a starting temperature reading before heating and decide on the range or type of measurements to be taken.

....

....

....

.... **(5 marks)**

(b) Suggest how you can determine the amount of thermal energy supplied to the heater by the electric current.

....

....

.... **(2 marks)**

(c) Explain how this experiment could be improved to give more accurate results.

....

....

.... **(2 marks)**

3 A known mass of ice is heated until it becomes steam. The temperature is recorded every minute. Describe how to use the data to identify when there are changes of state.

....

....

.... **(2 marks)**

4 Identify two hazards and subsequent safety measures that are common to both experiments to determine specific heat capacity and specific latent heat.

....

....

.... **(2 marks)**

Had a go ☐ Nearly there ☐ Nailed it! ☐

Pressure and temperature

1 Explain what is meant by temperature.

Consider the movement of particles.

.... **(1 mark)**

2 (a) Complete the table below showing some equivalent values in kelvin and degrees Celsius.

Kelvin (K)	degrees Celsius (°C)
	0
	−18
373	

(3 marks)

(b) (i) Describe what happens to a substance at the temperature absolute zero in terms of pressure and temperature.

....

....

....

....

.... **(3 marks)**

(ii) Give the value of the Celsius scale at which absolute zero occurs.

absolute zero = ... °C **(1 mark)**

3 In an experiment a fixed-volume container of 100 g of helium gas is warmed from −10 °C to 30 °C.

(a) Describe what happens to the velocity of the helium particles as a result of increasing temperature.

Guided

As the temperature increases ....

because .... **(2 marks)**

(b) Explain how this affects the pressure on the container walls.

....

....

.... **(2 marks)**

(c) State what happens to the average kinetic energy of the particles as the temperature increases.

.... **(1 mark)**

4 In a fixed volume of air the temperature in kelvin is increased by a factor of four. Explain how this affects the average kinetic energy of the air particles.

....

....

.... **(2 marks)**

Volume and pressure

1 Describe the pressure of a gas on a surface in terms of particle movement and force.

Guided

When particles of a gas collide with ....

they exert ....

.... **(3 marks)**

2 When a gas is at constant temperature, what is the relationship between volume and pressure?

☐ **A** Pressure is directly proportional to volume.

☐ **B** Volume and pressure both decrease.

☐ **C** Volume is inversely proportional to pressure.

☐ **D** Pressure and volume both increase. **(1 mark)**

3 The pressure in a cylinder of air is increased from atmospheric pressure (100 kPa) to 280 kPa. The original volume of the cylinder is 230 cm^3. Calculate the volume of the cylinder when the pressure is 280 kPa.

Rearrange the equation $P_1 \times V_1 = P_2 \times V_2$.

volume = ... cm^3 **(3 marks)**

4 A mixture of nitrous oxide and oxygen gases is used as an anaesthetic by dentists. 640 litres of the gas at normal atmospheric pressure (100 kPa) can be stored in a cylinder with an internal volume of 8 litres.

(a) Calculate the pressure needed to compress the gas into the cylinder.

pressure = ... Pa **(3 marks)**

(b) The gas is released from the cylinder at a rate of 2 litres per minute. Calculate the time the gas supply will last.

Guided

time = total volume ÷ rate

so time = ...

but 8 litres will be left in the cylinder at..

time = ... min **(2 marks)**

Extended response – Particle model

The transfer of thermal energy from a building may be reduced through the use of thermal insulation. Explain how insulation affects the transfer of thermal energy in various areas of a building.

You will be more successful in extended writing questions if you plan your answer before you start writing.

The question asks you to give a detailed explanation of thermal energy transferred out of a building and ways of reducing this. Think about:

- Important areas in a building from where thermal energy is transferred.
- The methods of thermal energy transfer through solids, fluids and a vacuum.
- Types of insulators and how they reduce the transfer of thermal energy.
- Examples of where to use insulators to reduce the transfer of thermal energy from a building.

You should try to use the information given in the question.

....

....

....

....

....

....

....

....

....

....

....

....

....

....

....

....

....

....

....

....

....

.... **(6 marks)**

Elastic and inelastic distortion

1 Draw a line from each force pair to the correct distortion it produces.

Force pair
push forces (towards each other)
pull forces (away from each other)
clockwise and anticlockwise

Distortion
stretching
bending
compression

(3 marks)

2 Give an example where each of the following may occur:

(a) tension

.... **(1 mark)**

(b) compression

.... **(1 mark)**

(c) elastic distortion

.... **(1 mark)**

(d) inelastic distortion.

.... **(1 mark)**

3 A student investigates loading two aluminium beams each with an elastic limit at 50 N. Beam 1 is tested to 45 N. Beam 2 is tested to 60 N. Predict what you would expect the beams to look like after the experiment. Explain your answer.

Guided

After testing, beam 1 would ....

....

Beam 2 would ....

.... **(4 marks)**

4 Car manufacturers use inelastic distortion to make cars safer. Discuss what they use and what the function of these items is.

> Think about how energy is absorbed to protect the passengers in the event of a crash.

....

....

....

....

.... **(3 marks)**

Had a go ☐ **Nearly there** ☐ **Nailed it!** ☐

Springs

1 State what is meant by the term elastic when describing an object experiencing a force.

....

....

.... **(2 marks)**

2 A spring is stretched from 0.03 m to 0.07 m, within its elastic limit. Calculate the force needed to stretch the spring. State the unit. Take the spring constant to be 80 N/m.

Guided

extension = 0.07 m − ....

force = ... × extension ....

....

force = ... unit .... **(3 marks)**

3 Deduce the spring constant that produces an extension of 0.04 m when a mass of 2 kg is suspended from a spring. Take *g* to be 10 N/kg.

Be careful not to confuse mass with force.

☐ **A** 0.02 N/m

☐ **B** 0.08 N/m

☐ **C** 50 N/m

☐ **D** 500 N/m **(1 mark)**

4 (a) Calculate the spring constant of a spring that is stretched 15 cm when a force of 30 N is applied.

k = ... N/m **(3 marks)**

(b) Calculate the energy transferred to the spring in (a).

You may find this equation useful
energy = ½ × spring constant × extension²

energy transferred = ... J **(2 marks)**

Forces and springs

1 (a) Describe how to set up an experiment to investigate the elastic potential energy stored in a spring using a spring, a ruler, masses or weights, a clamp and a stand.

Include a step to make sure the spring is not damaged during the experiment.

....

....

....

....

....

.... **(4 marks)**

(b) Explain why it is important to check that the spring is not damaged during the experiment.

....

....

.... **(2 marks)**

(c) Explain how the data collected must be processed before a graph can be plotted. Assume masses are used and measurements made in mm.

Guided

Masses must be converted to ....

The extension of the spring must be ....

....

Extension measurements should be .... **(3 marks)**

(d) Describe how a graph plotted from this experiment can be used to calculate:

(i) the elastic potential energy stored in the spring

.... **(1 mark)**

(ii) the spring constant k.

.... **(1 mark)**

(e) State the name of the law connecting the force, extension and spring constant.

.... **(1 mark)**

(f) Write the equation to calculate the energy stored by the spring.

.... **(1 mark)**

2 Explain the difference between the length of a spring and the extension of a spring.

....

.... **(1 mark)**

Had a go ☐ Nearly there ☐ Nailed it! ☐

Pressure and fluids

1 Explain what is meant by atmospheric pressure being 100 000 Pa.

....

....

....

.... **(2 marks)**

2 Explain why a diver experiences greater pressure from the water when she swims deeper.

....

....

....

.... **(2 marks)**

3 Calculate the water pressure on a submarine at a depth of 1500 m. Take the density of seawater to be 1025 kg/m^3 and g to be 10 N/kg.

You may find this equation useful
$P = h \times \rho \times g$

pressure = ... Pa **(2 marks)**

4 Calculate the total force bearing down on a football pitch due to atmospheric pressure. The football pitch is 100 m long and 50 m wide.

Guided

area of football pitch =

atmospheric pressure =

total force on pitch =

force = ... N **(4 marks)**

Upthrust and pressure

1 Which **two** of the following are units of pressure?

☐ **A** newtons (N)

☐ **B** newtons per metre^2 (N/m^2)

☐ **C** newtons per metre (N/m)

☐ **D** pascals (Pa) **(2 marks)**

2 A mooring buoy weighing 100 kg is floating in a river. State the upthrust on the buoy.

upthrust = ... N **(1 mark)**

3 Explain why a fully loaded container ship will float lower down in the water than when it has no containers on board.

....

.... **(2 marks)**

4 (a) Calculate the pressure on the ground of a 350 N block with an area of 0.3 m^2.

pressure = ... Pa **(2 marks)**

(b) Calculate the area of a block in contact with a table top. There is a force of 15 N normal to the table top exerting a pressure of 120 Pa.

area = ... m^2 **(2 marks)**

(c) Calculate the downwards pressure of a block with a bottom face of area 0.25 m^2 and a mass of 6 kg.

Take g = 10 N/kg.

pressure = ... Pa **(3 marks)**

5 Staff who work in historical houses often encourage visitors wearing stiletto-heeled shoes to replace them with flat slippers before touring old rooms with wooden floors. Explain why stiletto-heeled shoes could damage the floors but the slippers do not.

Guided

The stiletto shoe heel has a..

so the force per unit area of...than the ....

....

The larger pressure of .... **(4 marks)**

Had a go ☐ Nearly there ☐ Nailed it! ☐

Extended response – Forces and matter

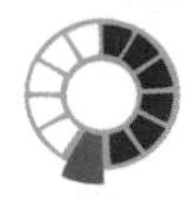

Explain the influence of weight, upthrust and density on a submarine when it floats on the surface of the sea, when it sinks and when it resurfaces.

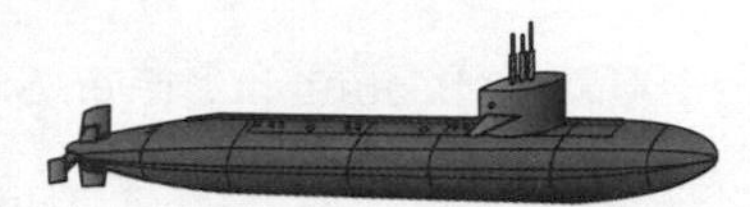

You will be more successful in extended writing questions if you plan your answer before you start writing.

The question asks you to give a detailed explanation of the how forces acting on a submarine change with its position in the water. Think about:

- How the displacement of water is linked to floating on the surface.
- The forces that acts upwards against the weight when the submarine is floating.
- The effect of the buoyancy force on the vertical position of the submarine.
- Why a submarine needs to change its overall density to sink.
- How a submarine can rise from deeper in the sea to the surface.

You should try to use the information given in the question.

.... **(6 marks)**

Timed Test 1

Time allowed: 1 hour 45 minutes

Total marks: 100

Edexcel publishes official Sample Assessment Material on its website. The Timed Test has been written to help you practise what you have learned and may not be representative of a real exam paper.

1 Charlie and Ravi plan to set up an experiment to measure speed. They have a trolley, an inclined ramp, a ruler and a stopwatch.

(a) Describe a method that the students could use to measure the speed of the trolley using the apparatus above. **(4 marks)**

(b) Suggest other apparatus the students could use to improve the precision of the data collected. **(2 marks)**

(c) Charlie and Ravi then extend their experiment to investigate the influence of another independent variable. The table below shows data collected by the students.

(cm)	(m)	(s)	
5	1.80	3.2	
10	1.80	2.4	
15	1.80	1.8	
20	1.80	1.4	
25	1.80	1.0	
30	1.80	0.4	
35	1.80	0.2	

(i) Deduce what the new independent variable might be and add titles for all three completed columns in the table. **(1 mark)**

(ii) Add a fourth column heading, with units, that could be added to the table to help with analysis. Add two correct entries for the fourth column. **(2 marks)**

(d) After travelling down the ramp, the trolley comes to a stop by itself. Discuss factors that would have caused the trolley to come to a stop and how these factors could be reduced. Your answer should refer to forces. **(4 marks)**

2 A crash-test car of mass 1000 kg is driven at the design testing centre to examine impact forces. The car starts from rest and accelerates to its final speed.

(a) Write the equation to calculate the acceleration of the car towards the crash barrier in a time *t*. **(1 mark)**

(b) The car is accelerated uniformly from 0 m/s to 10 m/s over a time of 20 s. Calculate how far the car will travel in m. **(2 marks)**

(c) Calculate the momentum of the car as it crashes into the crash barrier. State the unit. **(3 marks)**

(d) Which of the following would decrease the momentum in a collision?

☐ **A** Decreasing the crumple zones present.

☐ **B** Increasing the velocity of the vehicle.

☐ **C** Decreasing the mass of the vehicle.

☐ **D** Decreasing the time taken to change momentum. **(1 mark)**

Had a go ☐ **Nearly there** ☐ **Nailed it!** ☐

3 (a) A scientist is working in the countryside to investigate levels of radioactivity on a remote moor. A Geiger–Muller detector is used to measure the count rate. The level recorded is an average of 25 counts per minute.

(i) Give the name of the radiation the scientist is measuring. **(1 mark)**

(ii) State two sources of this radiation. **(2 marks)**

(iii) Suggest why the scientist takes a number of readings and then calculates an average. **(1 mark)**

(b) The scientist conducts a second experiment in the laboratory. First he measures the background radiation only. Next he tests a sample of radioactive material. The readings for background radiation are removed from those taken from the radioactive sample. The corrected count rate is recorded in a table.

Time (minutes)	Corrected count rate (counts per minute)
0	1030
1	760
2	515
3	387
4	258
5	194
6	129
7	98
8	65

(i) Plot a graph of the count rate against time. **(3 marks)**

(ii) Add a curve of best fit. **(1 mark)**

(iii) Determine the half-life in minutes of the radioactive source. **(1 mark)**

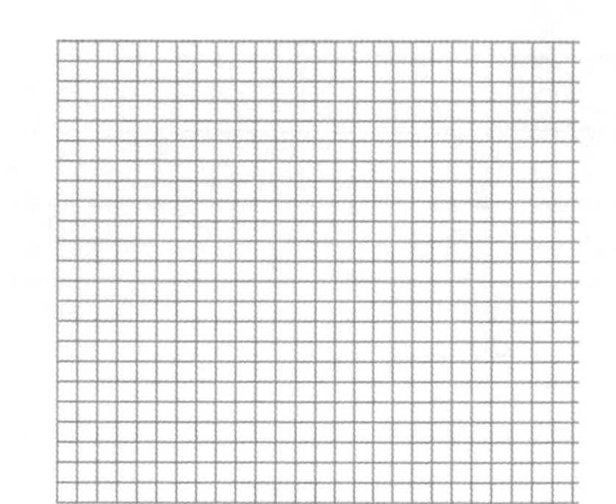

(c) Describe the nature of alpha, beta and gamma radiation in terms of ionising properties. **(3 marks)**

4 (a) Identify the stage that is not part of the life cycle of a star of similar mass to the Sun.

☐ **A** main sequence ☐ **C** supernova

☐ **B** nebula ☐ **D** white dwarf **(1 mark)**

(b) Describe the role of gravity in the life cycle of a star. **(4 marks)**

5 The diagram below shows the graph of a radio wave.

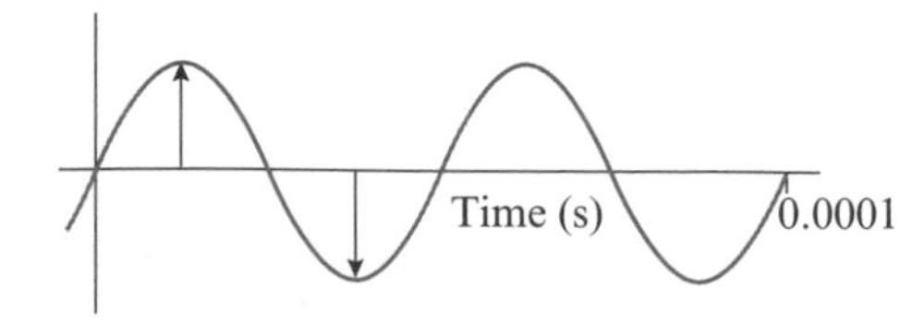

(a) (i) Write the equation linking wave speed, wavelength and frequency. **(1 mark)**

(ii) Identify the amplitude of the wave by adding A next to an arrow or marker on the graph. **(1 mark)**

(iii) Identify a wavelength of the wave by adding λ next to an arrow or marker on the graph. **(1 mark)**

(iv) State the meaning of wavelength in relation to your graph. **(2 marks)**

(v) Determine the frequency of the wave shown in the diagram in Hz. **(2 marks)**

(b) Sound waves can be used by fishing trawlers to locate shoals of fish.

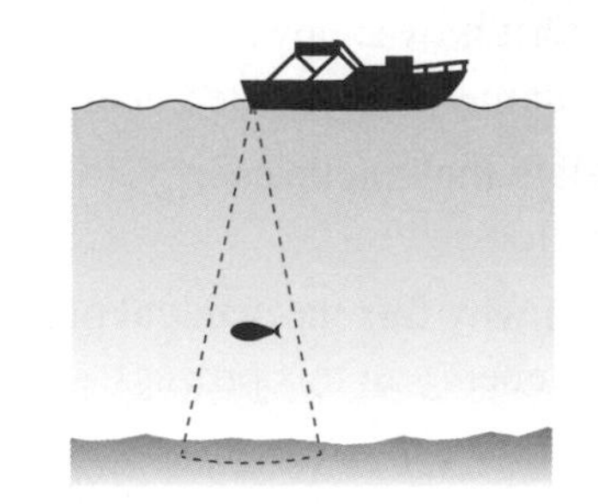

(i) State the name given to this process of using sound in water to determine the position of an object relative to a transmitter. **(1 mark)**

(ii) Explain how the ship can use sound waves to locate the shoal of fish. **(3 marks)**

6 The diagram below shows part of the electromagnetic spectrum.

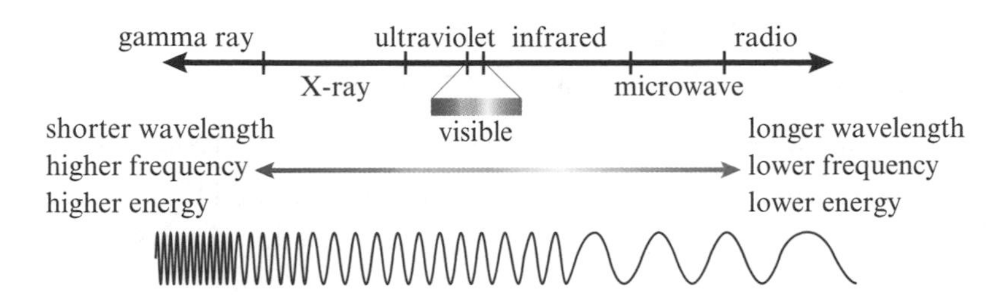

(a) State one use for each of the following waves:

(i) microwaves **(1 mark)**

(ii) ultraviolet **(1 mark)**

(iii) gamma rays. **(1 mark)**

(b) Explain which of the waves in (a) is the most damaging to body cells. **(2 marks)**

(c) Which **one** of the following terms best describes the nature of electromagnetic waves?

☐ **A** longitudinal ☐ **C** seismic

☐ **B** mechanical ☐ **D** transverse **(1 mark)**

(d) Light from the Sun takes 500 seconds to reach Earth, travelling at 3×10^8 m/s. Calculate the distance of Earth from the Sun. Give your answer in km. **(3 marks)**

(e) Explain why satellites are not used to transmit radio waves but are used to transmit microwaves. **(4 marks)**

7 (a) State the property of light that enables it to be used in fibre-optic communications. **(1 mark)**

(b) The diagram shows a light wave being refracted through a glass block.

glass
air
air
emergent ray
normal
incident ray

(i) Add the angle of incidence (i) and the angle of refraction (r) on the diagram. **(2 marks)**

(ii) Explain why the wave changes direction. **(2 marks)**

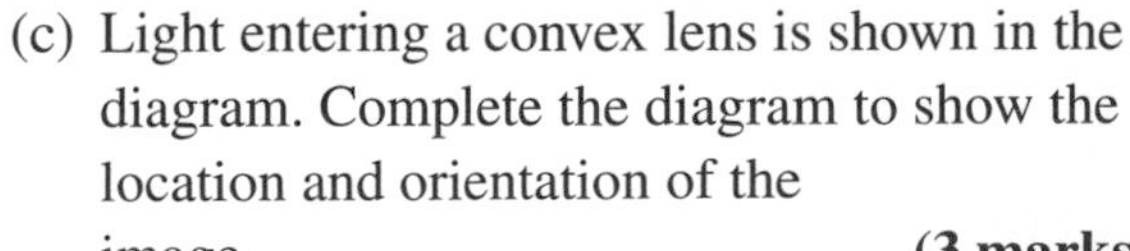

(c) Light entering a convex lens is shown in the diagram. Complete the diagram to show the location and orientation of the image. **(3 marks)**

object
F
F

8 A lift travels between floors in a building. The lift moves from the ground floor to the fourth floor through a height of 15 m in 20 s. The mass of the lift is 750 kg.

(a) (i) Calculate how much energy in J the lift gains in moving from the ground floor to the fourth floor. Take g to be 10 N/kg. **(2 marks)**

(ii) Calculate the average power that the lift exerts in moving the mass of 750 kg to the fourth floor. State the unit. **(3 marks)**

(b) As the lift moves upwards, not all of the energy supplied is usefully transferred. Suggest where some of the wasted energy is transferred to. **(2 marks)**

(c) Calculate the kinetic energy, in J, of the lift when it is moving from the ground floor to the fourth floor. **(4 marks)**

(d) Explain why this mechanical process could be described as wasteful. Give an example of wasted energy in this process. **(2 marks)**

9 (a) What is the missing colour in the partial sequence below of the colours in the visible light spectrum?

green → blue → .... → violet

☐ **A** cyan ☐ **B** indigo ☐ **C** purple ☐ **D** yellow **(1 mark)**

(b) At a music concert, the technician creates different colours of light to produce stage effects.

(i) Explain how coloured filters change the white spotlight to coloured lights. **(2 marks)**

(ii) One of the performers is wearing blue trousers and a red shirt. State what colour the clothes will appear in blue light. **(2 marks)**

(c) The technician fixes six new spotlights, each weighing 150 N and each with a flat base of 0.75 m^2, to the top of a 1 m-wide beam over the stage. Calculate the pressure in Pa the new spotlights add to the beam. **(3 marks)**

10 (a) In January 2016 five planets appeared to line up when viewed from Earth and were visible at the same time. These planets were Mercury, Venus, Mars, Jupiter and Saturn. Name the planets in our Solar System that were not part of this alignment. **(1 mark)**

(b) Most planets of the Solar System have natural satellites.

(i) Define the term natural satellite. **(1 mark)**

(ii) Give one example of a natural satellite. **(1 mark)**

(c) Explain why a satellite can have a constant speed but a changing velocity. **(2 marks)**

(d) Analysis of many of observations have shown the light from galaxies to be red-shifted. Explain why red-shift supports the theory of the expanding Universe. You should refer to absorption spectra in your answer. **(6 marks)**

Timed Test 2

Time allowed: 1 hour 45 minutes

Total marks: 100

Edexcel publishes official Sample Assessment Material on its website. This Timed Test has been written to help you practise what you have learned and may not be representative of a real exam paper.

1 A kettle is rated at 2000 W and is designed to operate on a 230 V mains supply.

(a) (i) Write the equation that links power, voltage and current. **(1 mark)**

(ii) Calculate the current drawn by the kettle in A. **(2 marks)**

(iii) Describe what could happen if a fuse were fitted that has too high or too low a rating. **(2 marks)**

(iv) What size of fuse should be fitted to the kettle to make it safe?

☐ **A** 1 A ☐ **B** 3 A ☐ **C** 10 A ☐ **D** 13 A **(1 mark)**

(b) (i) Give the equation that links energy, charge and potential difference. **(1 mark)**

(ii) State the relationship between current and charge. **(1 mark)**

(iii) State the unit for charge. **(1 mark)**

(c) Explain what happens to the energy of the system when the kettle is switched on. **(3 marks)**

2 The diagram shows a wire passing through a card.

(a) A compass is placed at various places on the card to determine the direction of the magnetic field around the wire. Explain whether the magnetic field is clockwise or anticlockwise, when viewed from above. **(3 marks)**

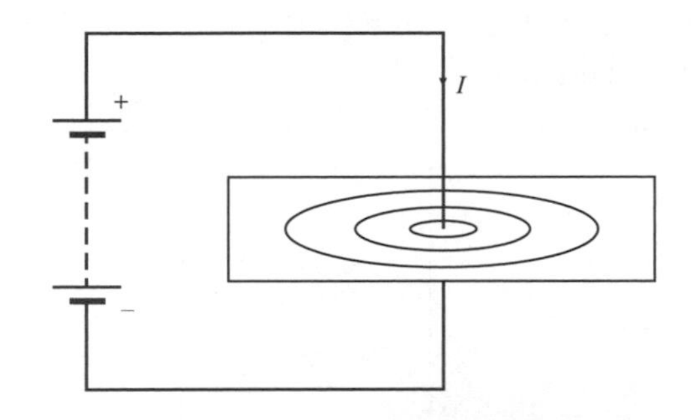

(b) (i) The wire is now turned into a solenoid. Draw the new magnetic field that this produces. **(4 marks)**

(ii) Suggest a use for this device. **(1 mark)**

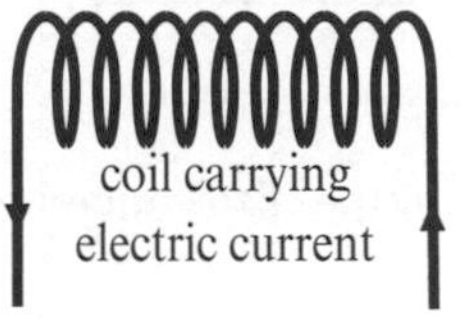

(c) A 0.75 m length of wire carrying a current of 3 A is placed between two magnets at right angles to the field of 0.5 T. Calculate the force experienced by the wire. Give the unit. **(3 marks)**

3 An electrical heater supplies electrical energy to a copper block of mass 2000 g at 12 V with a current of 12 A for 2 minutes.

(a) Calculate the energy supplied to the heater. State the unit. **(3 marks)**

(b) Calculate the temperature rise for the block of copper when supplied with 2880 J. The specific heat capacity of copper is 385 J/kg K. State the unit. **(3 marks)**

(c) Suggest why the actual temperature rise may be lower than the predicted value. **(1 mark)**

(d) (i) Explain how unwanted energy transfer in the heating of a metal block could be reduced. **(1 mark)**

(ii) Give an example of a suitable material that could be used. **(1 mark)**

4 A stationary fuel tanker is waiting on the runway to refuel a helicopter that has just landed and stopped close to the tanker. The helicopter has gained a static charge whilst travelling and is fitted with a metal rod, which touches the ground first when the helicopter lands.

(a) (i) Explain why the metal rod is necessary. **(2 marks)**

(ii) Suggest two reasons why this situation would be dangerous without the metal rod in place. **(2 marks)**

(b) Explain a second danger from static electricity that must be considered while the plane is being refuelled. **(3 marks)**

(c) Discuss how a thundercloud can also build up a significant static charge. Your answer should refer to the precautions people take to reduce the danger to buildings from lightning. **(6 marks)**

5 The circuit diagram shows three identical lamps and one cell.

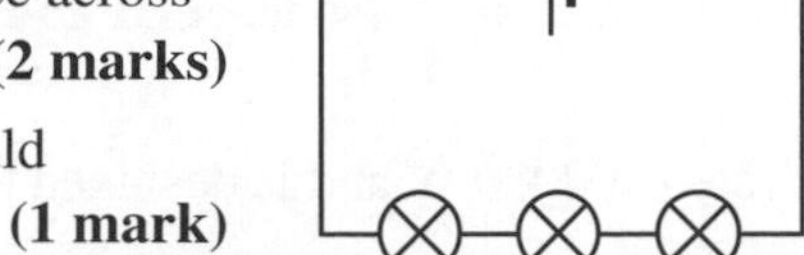

(a) The cell provides 1.5 V. Explain what the potential difference across each lamp is. **(2 marks)**

(b) Add a voltmeter to the circuit diagram to show how you could measure the potential difference across one lamp. **(1 mark)**

(c) Describe two ways of making the lamps brighter. **(2 marks)**

(d) The circuit diagram below shows a circuit to test a thermistor.

(i) Describe how resistance changes in a thermistor. **(3 marks)**

(ii) Give a use for a thermistor in a domestic circuit. **(1 mark)**

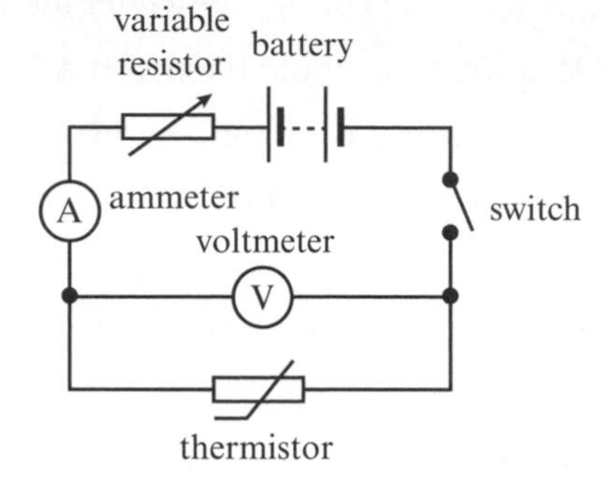

6 (a) Complete the boxes to show the differences in particle arrangement for elements of solids, liquids and gases. **(2 marks)**

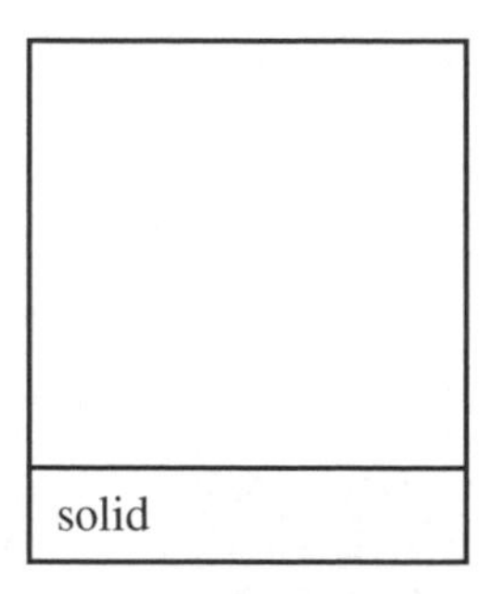

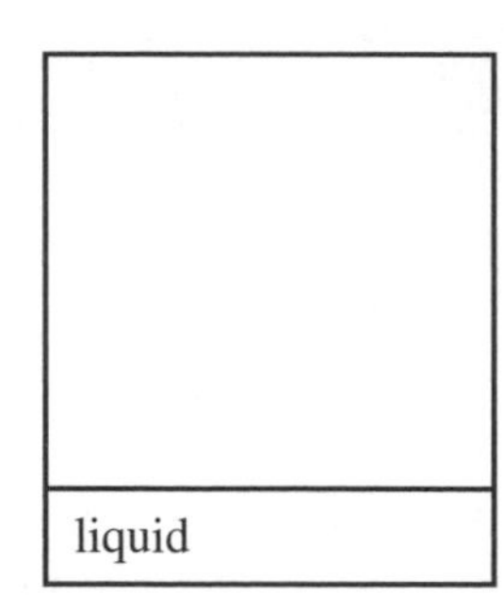

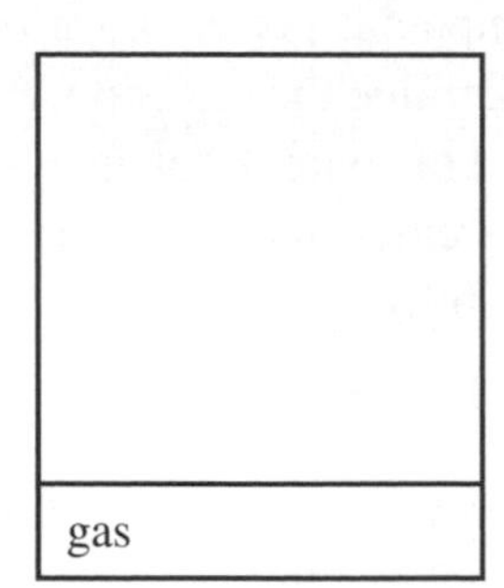

(b) Explain the difference between solids, liquids and gases in terms of the kinetic energy of particles. **(6 marks)**

7 (a) Write the equation that links density, mass and volume. **(1 mark)**

(b) A student conducts an investigation into density. She measures the dimensions of three rectangular blocks with a ruler to calculate the volume. Block 1 has a volume of 0.2 m^3 and a mass of 1500 g. Block 2 has a volume of 0.15 m^3 and a mass of 0.7 kg. Block 3 has the same volume as block 1 and the same mass as block 2. Deduce which block has the highest density.

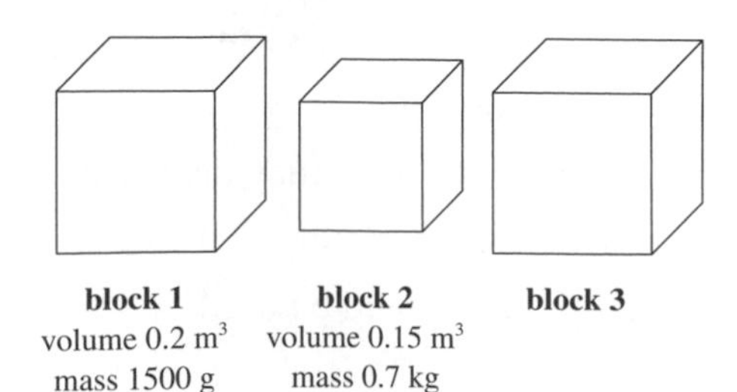

(4 marks)

(c) The student now investigates the density of three irregular rocks. She fills a eureka can with water, places the first rock into the can and collects the water displaced in a measuring cylinder.

(i) State what the student must do to make sure the reading on the measuring cylinder is as accurate as possible. **(2 marks)**

(ii) Describe how the student must make the other measurement to be able to calculate the density of the rock. **(2 marks)**

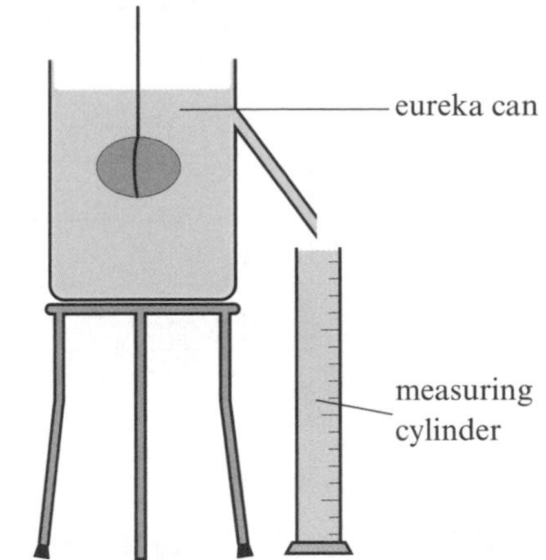

(d) The student is asked to display her results in a bar chart but she says that she cannot directly compare her calculated densities of the regular and irregular objects. Explain whether the student is right or wrong. **(2 marks)**

8 (a) Explain how step-up transformers improve the efficiency of energy transfers through the National Grid. **(2 marks)**

(b) The diagram shows a step-up transformer. Calculate the induced potential difference in the secondary coil when the primary potential difference is 250 V. Take n_p to be 150 and n_s to be 4500.

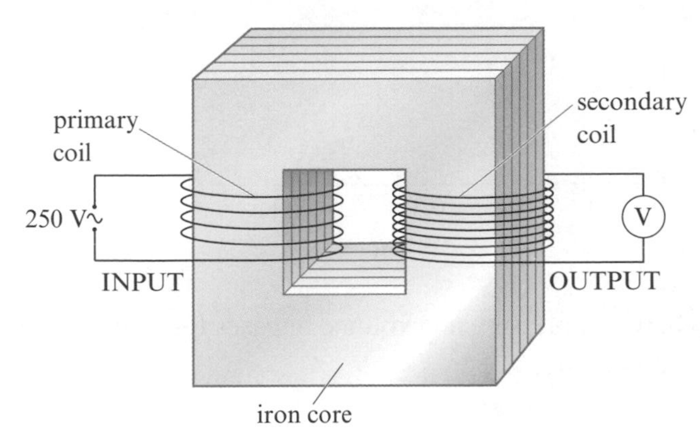

potential difference = ... V **(3 marks)**

(c) Explain what happens to the electricity supply before the electricity can enter a domestic building from the National Grid. **(2 marks)**

(d) (i) Calculate the current in the secondary coil of a step-down transformer, where $V_p = 4600$ V, $I_p = 5$ A and $V_s = 230$ V.

current = ... A **(3 marks)**

(ii) State the assumption made when calculating the power of a transformer. **(1 mark)**

9 Two students carry out an experiment to investigate Hooke's Law. They add masses to a spring by adding a 10 g mass after measuring each extension. They then measure the final extension. When the spring is unloaded the students find the spring has stretched.

(a) Name the point beyond which the spring will permanently change shape. **(1 mark)**

(b) Suggest two ways of improving the experiment. **(2 marks)**

(c) The table below shows results for the loading of a 32 mm spring.

Weight (N)	Length of spring (mm)
0	32
0.1	
	40
0.3	
0.4	48
	52
0.6	56

Complete the missing values in the table. **(2 marks)**

(d) Explain how the graph illustrates Hooke's Law. You should refer to the graph in your answer.

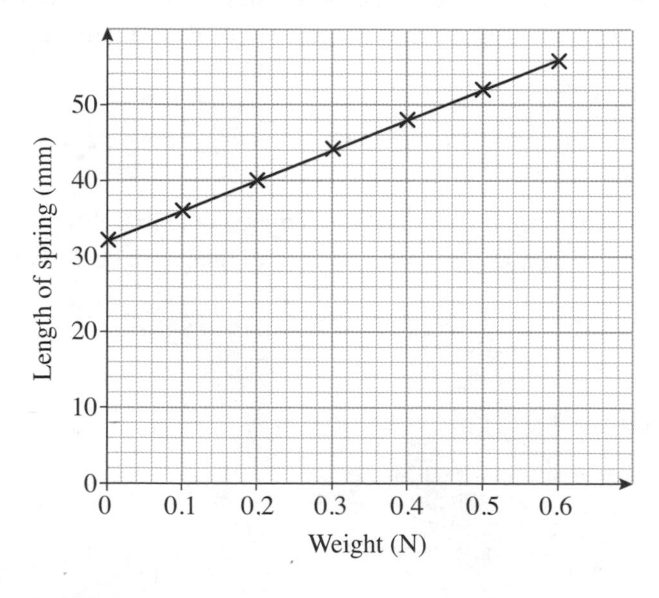

(4 marks)

10 (a) Identify the relationship between pressure and volume in gases for a constant amount of gas when held at a constant temperature.

☐ **A** pressure ∝ volume ☐ **C** pressure ∝ volume^2

☐ **B** pressure ∝ 1 ÷ volume ☐ **D** pressure ∝ volume^3 **(1 mark)**

(b) A bicycle is left outside in very cold weather overnight. In the morning the tyres appear not to be fully inflated. Later in the day when the temperature has risen significantly, the tyres look as though they have had air added, although this has not happened. Explain, with reference to the air particles inside the tyre, what caused the slight change in the appearance of the tyre. **(6 marks)**

Answers

Extended response questions

Answers to 6-mark questions are indicated with a star (*).
In your exam, your answers to 6-mark questions will be marked on how well you present and organise your response, not just on the scientific content. Your responses should contain most or all of the points given in the answers below, but you should also make sure that you show how the points link to each other, and structure your response in a clear and logical way.

1 Key concepts

1 All eight for **2 marks**, any four for **1 mark**: ampere, A; joule, J; pascal, Pa; coulomb, C; mole, mol; watt, W; newton, N; ohm, Ω.

2 A base unit is independent of any other unit **(1)**; a derived unit is made up from two or more base units. **(1)**

3 (a) 0.75 kg **(1)** (b) 750 W **(1)** (c) 1500 s **(1)**
(d) 0.03 m **(1)** (e) 3 000 000 J **(1)**

4 (a) 2.5 kHz = 2500 Hz **(1)**; 2.5 × 103 Hz **(1)**
(b) 8 nm = 0.000000008 m **(1)**; 8×10^{-9} m **(1)**

5 $s = d \div t$ = 75 m ÷ 10.5 s = 7.142 857 142 857 143 m/s **(1)** so to 5 significant figures = 7.1430 m/s **(1)**

2 Scalars and vectors

1 (a) scalars: speed, energy, temperature, mass, distance **(1)**; vectors: acceleration, displacement, force, velocity, momentum **(1)**
(b) (i) any valid choice and explanation, e.g. mass **(1)** is a scalar because it has a size/magnitude but no direction **(1)**
(ii) any valid choice and explanation, e.g. weight **(1)** is a vector because it is has a magnitude and direction **(1)**

2 (a) Velocity is used because both a size and a direction are given **(1)** and the swimmers are swimming in different directions. **(1)**
(b) The second swimmer is swimming in the opposite direction to the first swimmer. **(1)**

3 (a) D **(1)**
(b) Weight has a size/magnitude and a direction but all the other quantities just have a magnitude. **(1)**

4 Both the wind speed and the speed of the aeroplane have direction as well as magnitude and so are vectors. **(1)**. As they have different directions, **(1)**, the pilot needs to take this into account when planning the route and this will also affect the time taken to fly the rout. **(1)**

3 Speed, distance and time

1 (a) (i) B **(1)**
(ii) C **(1)**
(b) In part A he travels 60 m in 40 s. Speed = distance ÷ time **(1)** = 60 m ÷ 40 s **(1)** = 1.5 m/s **(1)** (You can use any part of the graph to read off the distance and the time as the line is straight; you should always get the same speed).
(c) Displacement is the length and direction of a straight line between the runner's home and the park **(1)** but the distance the runner ran probably included bends and corners on the path the runner took. **(1)**

2 (a) speed = 84 m ÷ 24 s **(1)** = 3.5 **(1)** m/s **(1)**
(b) 3.5 m/s upwards or up **(1)**

3 time = distance ÷ speed = 400 m ÷ 5 s **(1)** = 80 s **(1)**

4 Equations of motion

1 All four correct for **2 marks**; two or three correct for **1 mark**. a: acceleration; t: time; x: distance; u: initial velocity

2 (a) $a = (v - u) \div t$ **(1)** = (25 m/s – 15 m/s) ÷ 8 s **(1)** = 1.25 m/s^2 **(1)**
(b) $v^2 = u^2 + 2 \times a \times x$ = (25 m/s)^2 + 2(1.25 m/s^2 × 300 m) **(1)** = 1375 m^2/s^2 **(1)**
$v = \sqrt{1375}$ m/s = 37 m/s **(1)**
(c) (allow rounding error: answers between 37.00 m/s and 37.10 m/s) $v^2 - u^2 = 2 \times a \times x$ so $x = (v^2 - u^2) \div (2 \times a)$ **(1)** = ((5 m/s)^2 – 1375 m^2/s^2) ÷ (2 × –2 m/s^2) **(1)** = –1350 m^2/s^2 ÷ –4 m/s^2 = 337.5 m **(1)**

5 Velocity/time graphs

1 (a) $a = (v - u) \div t$ = (4 – 0) ÷ 5 **(1)** = 0.8 **(1)** m/s^2 **(1)**
(b) Plot a velocity/time graph **(1)**; the total area under the line can be calculated **(1)** and this gives the total distance travelled. **(1)**

2 (a) A, **(1)** C **(1)**
(b) a right-angled triangle with a horizontal and vertical side that covers as much of the line as possible **(1)**
(c) change in velocity = 30 m/s, time taken for change = 5 s; acceleration = (change in velocity) ÷ (time taken) = 30 m/s ÷ 5 s **(1)** = 6 m/s^2 **(1)**; (the triangle drawn may be different but the answer should be the same);
(d) area under line = ½ × 5 s × 30 m/s **(1)** = 75 m **(1)**

6 Determining speed

1 Commuter train: 55 m/s; speed of sound in air: 330 m/s; walking: 1.5 m/s **(1)**. All three needed for **1 mark**.

2 (a) The light beam is cut/broken by the card as it enters the light gate, **(1)**, when the card has passed through the light beam is restored **(1)** and this stops the timer. **(1)**
(b) Speed is calculated from the length of the card and the time taken for the card to pass through the light gate. **(1)**

3 Very short distances may be measured using this method and this gives a good measure of instantaneous speed **(1)**. Short times are difficult to measure accurately with a stopwatch **(1)**. Errors associated with human error/reaction time/parallax are reduced. **(1)**

7 Newton's first law

1 Four arrows drawn: vertical: down = weight, up = upthrust (arrows the same length); horizontal left to right = thrust or force from the engines, right to left = water resistance or drag, the thrust arrow should be longer than the drag arrow (**1 mark** for all the forces correctly named and **1 mark** for the corresponding relative lengths of the arrows).

2 (a) The action is the downwards force of the skater on the ice **(1)** and the reaction is the (upwards) force of the ice on the skater. **(1)** (Can be the other way around).
(b) resultant force = 30 N – 10 N – 1 N **(1)** = 19 N **(1)**
(c) The resultant force is zero/0 N **(1)** so the velocity is constant/stays the same. **(1)**
(d) Only the resistance forces are acting now, so the resultant force is backwards/against the motion of the skater **(1)**, so the skater slows down/has negative acceleration. **(1)**

3 (a) Assume downwards is positive; so resultant downward force = 1700 N – 1900 N = – **(1)** 200 N **(1)** (State which direction you are using as the positive direction.)
(b) The velocity of the probe towards the Moon will decrease **(1)** because the force produces an upwards acceleration/negative acceleration. **(1)**

8 Newton's second law

1 (a) The trolley will accelerate **(1)** in the direction of the pull/force. **(1)**
(b) The acceleration is smaller/lower **(1)** because the mass is larger. **(1)**

2 (a) $F = m \times a$ = 3000 N × –13 m/s^2 **(1)** = –39 000 **(1)** N **(1)**;
(b) in the opposite direction to the spacecraft's motion/upwards **(1)**

3 (a) $a = F \div m$ = 10 500 N ÷ 640 kg **(1)** = 16.4 **(1)** m/s^2 **(1)**
(b) The mass of the car decreases **(1)** so the acceleration will increase. **(1)**

9 Weight and mass

1 The mass of the LRV on the Moon is 210 kg **(1)** because the mass of an object does not change if nothing is added or removed. **(1)**

2 B **(1)**

3 $W = m \times g$ **(1)** so (1 + 2 + 1.5) kg × 10 N/kg = 4.5 kg × 10 N/kg **(1)** = 45 N **(1)**

4 calculating correct masses for all three items **(1)**; selecting correct items **(1)**; clothes 10.5 kg + camera bag 5.5 kg + jacket 3.5 kg = 19.5 kg **(1)**

10 Force and acceleration

1 Electronic equipment is much more accurate **(1)** than trying to obtain accurate values for distance and time to calculate velocity, then calculate acceleration, **(1)** when using a ruler and a stopwatch. (Reference should be made to distance, time and velocity.)

2 Acceleration is inversely proportional to mass. **(1)**

3 Acceleration is the change in speed ÷ time taken so two velocity values are needed **(1)**; the time difference between these readings **(1)** is used to obtain a value for the acceleration of the trolley.

4 (a) For a constant slope, as the mass increases, the acceleration will decrease **(1)** due to greater inertial mass. **(1)**
(b) Newton's second law, $a = F \div m$ **(1)**

5 An accelerating mass of greater than a few hundred grams can be dangerous and may hurt somebody if it hits them at speed. **(1)** Any two of the following precautions: do not use masses greater than a few hundred grams **(1)**, wear eye protection **(1)**, use electrically tested electronic equipment **(1)**, avoid trailing electrical leads. **(1)**

11 Circular motion

1 (a) The velocity of an orbiting satellite changes because the direction is constantly changing **(1)** even though speed remains constant. **(1)**
(b) The force provided by the Earth's gravitational field causes it to change direction as it orbits the Earth at constant speed. **(1)**

2 B **(1)**

3 tension, **(1)** any valid example of a rope/string and mass, etc. **(1)**; gravitational, **(1)** any valid example of an orbiting mass **(1)**; frictional, **(1)** any circling body with dynamic friction **(1)**

4 (a) centripetal force **(1)**
(b) The ball is constantly changing direction **(1)** (due to the centripetal force acting on it).

12 Momentum and force

1 B **(1)**

2 (a) Force is the rate of change of momentum **(1)**. It is the change in momentum divided by the time take for the change. **(1)**
(b) $p = m \times v$ **(1)** = 1500 kg × 25 m/s **(1)** = 37 500 kg m/s, so change in momentum = 37 500 kg m/s **(1)**
(c) The forces exerted on the passenger are large when the mass is large **(1)** or the deceleration of the vehicle is large **(1)**; an airbag/crumple zone/seat belt **(1)** increases the time over which a passenger comes to rest so reducing the force exerted on them. **(1)**

3 $F = (mv - mu) \div t$ = ((500 kg × 15 m/s) – (500 kg × 10 m/s)) **(1)** ÷ 20 s = (7500 – 5000) kg m/s ÷ 20 s **(1)** = 125 N **(1)**

4 Any three from: The hockey player should try to make sure that the change in speed between the hockey stick and the ball is as high as possible/move the hockey stick very quickly **(1)**. The contact time between the club and ball needs to be as small as possible/ the ball needs to be hit quickly **(1)**. The large change in velocity **(1)** and the small contact time **(1)** make the force to move the ball as large as possible. **(1)**

13 Newton's third law

1 D **(1)**

2 momentum = mass × velocity **(1)**

3 momentum = 1200 kg × 30 m/s **(1)**; momentum = 36 000 kg m/s **(1)** in the south direction. **(1)**

4 (a) momentum of Dima and car = 900 kg × 1.5 m/s = 1350 **(1)** kg m/s
(b) (i) momentum of Sam and car = 900 kg × 3 m/s = 2700 **(1)** kg m/s
(ii) The total momentum of the two cars after the collision must equal the total momentum of the two cars before the collision **(1)** so total momentum of both cars is unchanged. **(1)**
(iii) total momentum = 1350 + 2700 = 4050 kg m/s **(1)** so velocity after collision = total momentum ÷ total mass = 4050 kg m/s ÷ 1800 kg **(1)** = 2.25 m/s **(1)**

5 momentum of skater 1 before collision = 50 kg × 7.2 m/s = 360 kg m/s; momentum of skater 2 before collision = 70 kg × 0 m/s = 0 kg m/s; momentum of both skaters after collision = 360 + 0 = 360 kg m/s **(1)** so combined velocity = 360 kg m/s ÷ (70 + 50) kg **(1)** = 3 m/s **(1)**

14 Human reaction time

1 B **(1)**

2 Human reaction time is the time between a stimulus occurring and a response **(1)**. It is related to how quickly the human brain can process information and react to it. **(1)**

3 A person sits with their index finger and thumb opened to a gap of about 8 cm **(1)**. A metre ruler is held, by a partner, so that it is vertical and exactly level with the person's finger and thumb, with the lowest numbers on the ruler at the bottom **(1)**. The ruler is dropped and then grasped by the other person. **(1)**

4 (a) 0.20–0.25 s **(1)**
(b) Success in certain professions relies on short reaction times, where a fast response could result in ensuring the safety of others **(1)** or is necessary for a competitive career, such as sports **(1)**. Any two suitable examples with justification, e.g. an international tennis player who is able to react to the opponent hitting the ball **(1)** or a racing driver who can react quickly to changing situations. **(1)**

5 Time = distance ÷ speed so time = 25 ÷ 20 **(1)** Reaction time = 1.25 **(1)** (s).

15 Stopping distance

1 (a) thinking distance + braking distance = overall stopping distance **(1)**
(b) speed increases by 3 times so thinking distance increases by 3 so thinking distance = 3 × 6 m = 18 m **(1)**; speed increases by 3 times so braking distance increases by 9 so braking distance = 9 × 6 m = 54 m **(1)**; total stopping distance = 18 m + 54 m = 72 m **(1)**
(c) Thinking distance will increase if: the car's speed increases, the driver is tired, the driver has taken alcohol or drugs **(1)** (all three needed for **1 mark**). Braking distance will increase if: the road is icy or wet, the brakes or tyres are worn, the mass of the car is bigger. **(1)** (All three needed for **1 mark**).
(d) Worn tyres will have less surface area in contact with the road so frictional force opposing motion will be less **(1)** therefore the car will take longer to stop than when the tyres were new. **(1)** (This affects the braking distance component. The thinking distance component remains unchanged).

2 $F \times d = ½\, m \times v^2$, so $F = ½\, m \times v^2 \div d$, = ½ × 1500 kg × (8 m/s × 8 m/s) ÷ 75 m, **(1)** 750 kg × 64 m²/s² ÷ 75 m **(1)** = 640 **(1)** N

3 Driving faster will increase thinking distance **(1)** and braking distance **(1)**. If drivers do not increase their normal distance behind the vehicle in front accordingly there is an increased risk of an accident/collision. **(1)**

16 Extended response – Motion and forces

*Answer could include the following points.

- Acceleration is the rate of change of velocity (speed in a given direction) so although the speed is constant, the direction is continually changing for an object in circular motion.
- For motion in a circle there must be a resultant force known as a centripetal force that acts towards the centre of the circle.
- The string provides the centripetal force which acts towards the centre of the circle.
- Extend the investigation with different lengths of string.
- Extend the investigation with different masses.
- Improve data collection with electronic sensors.
- Improve data collection with video analysis.
- Reference to the importance of control variables for valid data collection.

17 Energy stores and transfers

1 B **(1)**

2 The energy transfer diagram shows that the total amount of energy in the stores before the transfer is equal to the total amount of energy in the stores after the transfer **(1)** so there is no net change, supporting the conservation of energy. **(1)**

3 chemical store **(1)**; kinetic store and thermal store **(1)**; thermal store **(1)**

4 (a) Gravitational store bar is lower **(1)**; kinetic store bar higher **(1)**; rise in kinetic store bar should equal the amount down in the gravitational store bar. **(1)**
(b) Chemical store bar is lower **(1)** than kinetic store bar **(1)**; thermal store bar higher **(1)**; total heights of all bars should be the same as the total heights of the bars before use. **(1)**

18 Efficient heat transfer

1 concrete **(1)**: this has the lowest relative thermal conductivity, which means it will have the slowest rate of transfer of thermal energy. **(1)**

2 (a) Thicker walls provide more material for the thermal energy to travel through from the inside to the outside **(1)** so the rate of thermal energy loss is less, keeping the houses warmer. **(1)**
(b) Thicker walls provide more material for the thermal energy to travel through from the outside to the inside **(1)** so the rate of

thermal energy transfer is less, keeping the houses cool. **(1)**

3 useful energy transferred = energy transferred to the box = 1 000 000 J; total energy used by the crane = the energy stored in the fuel = 4 000 000 J; efficiency = 1 000 000 J ÷ 4 000 000 J 100 **(1)** = 25% **(1)** (or calculation can omit × 100 and leave efficiency as 0.25)

4 (a) efficiency = 20% ; thermal (wasted) energy = 80% **(1)** so this is 4 × 40 = 160 J **(1)**; so total energy in = 40 + 160 = 200 J **(1)**
(b) 200 W **(1)**

19 Energy resources

1 (a) geothermal **(1)**
(b) Demand is greatest at certain times of the day **(1)**. Demand may be high when some renewable sources may not be available. **(1)**

2 (a) A hydroelectric power station is a reliable producer of electricity because it uses the gravitational potential energy of water which can be stored until it is needed **(1)**. As long as there is no prolonged drought/ lack of rain the supply should be constant. **(1)**
(b) Any of the following Hydroelectric power stations have to be built in mountainous areas/high up (compared to supply areas so that the gravitational potential energy can be captured) **(1)**. The UK has very few mountainous areas like this **(1)**. It is limited to areas such as north Wales and Scottish Highlands. **(1)**

3 (a) When carbon dioxide is released into the atmosphere it contributes to the greenhouse effect/build-up of CO_2 **(1)** which is believed to contribute to global warming. **(1)**
(b) Sulfur dioxide and nitrogen oxides have been found to dissolve in the water droplets in rain clouds, increasing their acidity **(1)**; this can kill plants/damage forests and lakes/dissolve the surfaces of historical limestone buildings. **(1)**
(c) Fossil fuel power stations do not rely on the energy stores in the environment **(1)** and so can be built in re-developed areas/do not need to be positioned to take advantage of wind/tidal/wave/ hydroelectric resources. **(1)**

4 Any two from: We do not know how much oil there is to extract **(1)**, how fast it can be extracted **(1)** or how demand for oil may change. **(1)**

20 Patterns of energy use

1 (a) After 1900 the world's energy demand rose as the population grew **(1)**. There was development in industry which increased the demand for energy **(1)** and the rise of power stations using fossil fuels added to demand. **(1)**
(b) (i) Non-renewable energy resources / fossil fuels **(1)**,
(ii) Any two from; population has increased so energy consumption is higher **(1)**, industrial/ technological developments require more energy **(1)**, transport networks have grown. **(1)** (any other valid reason)
(iii) Nuclear research began in the 1940s. **(1)**
(iv) hydroelectric **(1)**

2 Any six points made for **6 marks**: As the population continues to rise the demand for energy will also continue to rise. **(1)** Current trends show the use of fossil fuels being a major contributor to the world's energy resources **(1)**. These are running out and no other energy resource has so far taken their place **(1)**. This could lead to a large gap between demand and supply **(1)**. To match the rise in demand for energy, further research and development of non-renewable resources will need to be made **(1)** to provide reliable **(1)** and cost-effective **(1)** energy supplies. While cheaper fossils fuels still remain available there is less incentive for governments to do this. **(1)**

21 Potential and kinetic energy

1 D **(1)**

2 Kinetic energy = ½ × m × v^2 **(1)** = ½ × 70 × 6^2 **(1)** = 1260 **(1)** J **(1)**

3 (a) ΔGPE = 2000 kg × 10 N/kg × 0.5 m **(1)** = 10 000 J **(1)**
(b) 10 000 J (or same answer as given to part (a) **(1)**
(c) KE = work done or GPE gained = 10 000 J **(1)**; v^2 = KE ÷ 0.5 × m **(1)** so v^2 = 10 000 ÷ 1 000 **(1)** = 10 so v = $\sqrt{10}$ = 3.16 m/s **(1)**

4 *The indicative content below is not prescriptive and you are not required to include all the material which is indicated as relevant. Additional content included in the response must be scientific and relevant.*

Four from: Kinetic energy **(1)** transferred to the ball reduces **(1)** as it climbs to the top of the curve where KE is a minimum **(1)** and gravitational energy reaches maximum **(1)**. Some of the kinetic energy transferred to the ball is dissipated to the surroundings **(1)** as thermal energy **(1)** due to air resistance/drag/ friction. **(1)**

22 Extended response – Conservation of energy

*Answer could include the following points.

- Refer to the change in gravitational potential energy (GPE) as the swing seat is pulled back/raised higher.
- Before release, the GPE is at maximum/ kinetic energy (KE) of the swing is at a minimum.
- When the swing is released, the GPE store falls and the KE store increases.
- KE is at a maximum at the mid-point, GPE is at a minimum.
- The system is not 100% efficient; some energy is dissipated to the environment.
- Friction due to air resistance and/or at the pivot results in the transfer of thermal energy to the surroundings/environment.
- Damping, due to friction, will result in the KE being transferred to the thermal energy store of the swing and hence to the environment.
- Eventually all the GPE will have been dissipated to the surroundings/environment (so is no longer useful).

23 Waves

1 Sound waves are this type of wave: L; All electromagnetic waves are this type of wave: T; Particles oscillate in the same direction as the wave: L; They have amplitude, wavelength and frequency: B; Seismic S waves are this type of wave: T; They transfer energy: B. All six correct – **3 marks**; five correct – **2 marks**; three correct – **1 mark**.

2 (a) B **(1)**
(b) 6 cm/0.06 m **(1)**
(c) Any correct wave with higher amplitude **(1)** and shorter wavelength **(1)**.

3 (a) When a sound wave is generated each particle oscillates **(1)** in the same direction as the direction in which the wave travels. **(1)**
(b) When a water wave is generated the surface particles oscillate **(1)** at 90°/ perpendicular to the direction in which the wave travels. **(1)**

24 Wave equations

1 distance travelled by the waves (in metres) = 30 000 m **(1)**; time taken = 20 s; speed of sound = 30 000 m ÷ 20 s **(1)** = 1500 m/s **(1)**

2 (a) $v = f \times \lambda$ or wave speed = frequency × wavelength) **(1)**
(b) wave speed = 0.017 m × 20 000 Hz **(1)** = 340 m/s **(1)**

3 $\lambda = v \div f$ = 0.05 m/s ÷ 2 **(1)** = 0.025 **(1)** m **(1)**

4 $x = v \times t$ **(1)** = 300 000 000 m/s × 0.12 s **(1)** = 36 000 km **(1)**

25 Measuring wave velocity

1 frequency of the waves (f) = 3 Hz; wavelength of the waves (λ) = 0.05 m; speed of waves = 3 Hz × 0.05 m **(1)** = 0.15 **(1)** m/s **(1)**

2 D **(1)**

3 T = 4 divisions × 0.005 s/division = 0.02 s **(1)**; $f = 1 \div T$ = 1 ÷ 0.02 s **(1)** = 50 Hz **(1)**

4 (a) They can process the data using the equation $v = x \div t$ **(1)**
(b) Use an electronic data collector **(1)**; repeat the experiment at 50 m **(1)**; repeat the experiment over a range of distances. **(1)**

26 Waves and boundaries

1 REFLECTION: the wave bounces back at a surface but does not pass through. REFRACTION: the wave passes through but at a changed speed. ABSORPTION: the wave energy is transferred into a thermal energy store. All three correct for **2 marks**; two correct for **1 mark**.

2 (a) C **(1)**
(b) Stone is the most dense **(1)** and the greater the difference in density between materials, the more sound energy will be reflected. **(1)**

3 In warm air, the particles have more kinetic energy **(1)**. As they move faster they collide with greater frequency **(1)** and carry the waves more quickly **(1)**, vibrations are transferred more easily. **(1)**

4 The greater the difference in density **(1)**, the more sound energy will be reflected **(1)**. Sound is transmitted from one material to

another **(1)** when their densities are similar **(1)**. Any two valid examples, e.g. reflection between air and concrete **(1)** or transmission of sound through a plaster/brick wall. **(1)**

27 Sound waves and the ear

1 B **(1)**

2 The length of the string determines the frequency of the note **(1)**. Each string produces a different note. **(1)**

3 The particles are much further apart in a gas than in a solid **(1)** so it is much more difficult for vibrations to be passed from one particle to another. **(1)**

4 In the human ear the ear drum will not vibrate **(1)** if the wave frequency is less than 20 Hz or more than 20 kHz (range must be stated) **(1)**. If there is no vibration of the ear drum, no sound is heard. **(1)**

28 Ultrasound and infrasound

1 (a) Infrasound is sound that has a frequency lower than 20 Hz. **(1)**
(b) C **(1)**
(c) Scientists set off explosions **(1)** and use detectors **(1)** to receive the infrasound waves. These are then used to understand rocks/locate petroleum. **(1)**

2 $d = s \times t$ = 1500 m/s × 4 s = 6000 m **(1)** but this is the return sound so the depth will be half this, so depth = 6000 m ÷ 2 = 3000 m **(1)**

3 As ultrasound waves pass into the body some waves are reflected each time they meet a layer of tissue **(1)** with a different density **(1)**. The scanner detects the echoes **(1)** and the computer uses the information to make a picture. **(1)**

4 distance travelled by ultrasound = speed × time **(1)** = 8400 m/s × 0.5 × 10^{-9} s **(1)** = 4.2 × 10^{-6} m **(1)**; distance of layer below the surface = ½ × 4.2 × 10^{-6} = 2.1 × 10^{-6} m **(1)**

5 Infrasound provides information that could not normally be detected **(1)** so by monitoring seismic activity possible dangers can be detected **(1)** and the population warned/evacuated. **(1)**

29 Sound wave calculations

1 Two sets of suitable numbers for wavelength **(1)** and frequency **(1)** should be inserted into the wave equation $v = f \times \lambda$ to illustrate the inversely proportional relationship; e.g. 200 = 4 × 50 **(1)** and 200 = 2 × 100 **(1)**

2 (a) As sound waves pass from a low-density material to a high-density material the speed of sound increases. **(1)**
(b) The denser the material, the greater the speed **(1)**, since the energy can be passed more easily in a dense material from particle to particle. **(1)**

3 (a) The Earth's mantle becomes more dense with increasing depth **(1)**. The wave speed depends on the density **(1)**, which in turn depends on the increasing pressure. **(1)**
(b) Waves are refracted when they meet a boundary of different densities **(1)** so this tells us that the Earth has layers of different densities. **(1)**
(c) $\lambda = v \div f$ = 7 m/s ÷ 0.05 Hz **(1)** = 140 m **(1)**

30 Waves in fluids

1 (a) Count the number of waves that pass a point each second and do this for one minute **(1)**; divide the total by 60 to get a more accurate value for the frequency of the water waves. **(1)**
(b) Use a stroboscope to 'freeze' the waves **(1)** and find their wavelength by using a ruler in the tank/on a projection. **(1)**
(c) wave speed = frequency × wavelength or $v = f \times \lambda$ **(1)**
(d) the depth of the water **(1)**

2 A ripple tank can be used to determine a value for the wavelength, frequency and wave speed of water waves **(1)**, as long as small wavelengths **(1)** and small frequencies are used. **(1)**

3 water: hazard – spills may cause slippages; safety measure – report and wipe up immediately **(1)**; electricity: hazard – may cause shock or trailing cables; safety measure – do not touch plugs/wires/switches with wet hands or keep cables tidy **(1)**; strobe lamp: hazard – flashing lights may cause dizziness or fits; safety measure – check that those present are not affected by flashing lights **(1)**

31 Extended response – Waves

*Answer could include the following points.

- The phenomena shown are refraction (from the key to the driver) and reflection (from the Sun to the key). These are properties of waves.
- Light is a wave and so is refracted through transparent or translucent materials.
- Light is a wave and so can be reflected at an opaque surface.
- Light waves from the Sun are reflected by the key.
- When light passes from one material to another of different density it is refracted at the boundary due to a change of speed.
- When light passes from a more dense material to a less dense material it is refracted away from the normal (as in this case).
- The actual position of the key is at position A.
- The key appears to be at position B because the man's brain extrapolates the refracted wave (as shown by the dashed line).

32 Reflection and refraction

1 angle of incidence (i) = angle of reflection (r) **(1)**

2

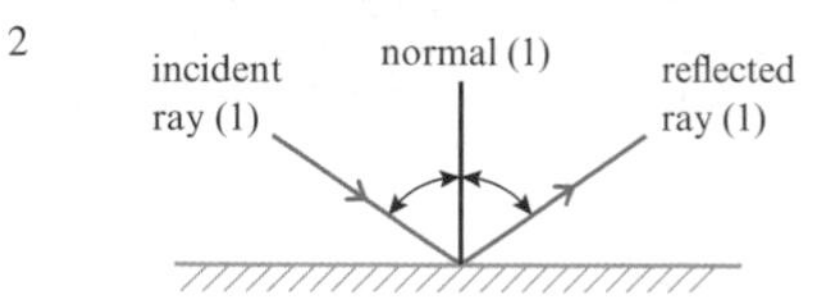

angle of incidence and angle of reflection (1)

3 The ray of light bends away from the normal. **(1)**

4 (a) Wave fronts drawn in the glass **(1)**; direction of travel of wave fronts is closer to normal **(1)**; wave fronts are closer together (shorter wavelength. **(1)**
(b) The speed of the wave/light changes **(1)** and this changes the direction of the wave. **(1)**

33 Total internal reflection

1 (a) (i) The light slows down **(1)** as it passes from a less dense to a more dense material. **(1)**
(ii) The light speeds up **(1)** as it passes from a more dense to a less dense material. **(1)**
(b) When light enters a glass block at an angle of zero degrees to the normal/along the normal line/at 90° to the boundary/interface. **(1)**

2 (a)

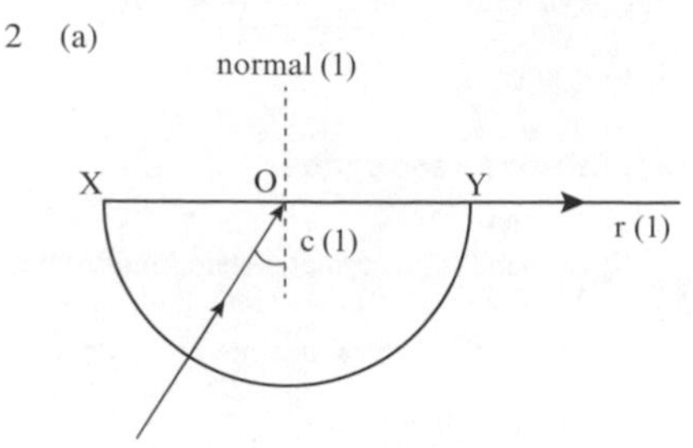

3 (a) Diagram must show at least three equal angles of internal reflection (by eye). **(2)**
(b) The cladding must have a lower refractive index than the core. **(1)**

4 Any four from: Endoscopes **(1)** use total internal reflection **(1)** in a bunch of optical fibres **(1)** to look inside patients' bodies. **(1)** This allows only small cuts to be made, promoting faster healing time/no cuts at all. **(1)**

34 Colour of an object

1 (a) Any three valid examples, e.g. radio waves reflected from the ionosphere **(1)**, light reflected through a periscope **(1)**, infrared waves reflected on the surface of a solar furnace. **(1)**
(b) any surface such as a mirror, smooth water, glass **(1)**

2 violet: shortest wavelength of visible light **(1)**; yellow: absorbed by a red object **(1)**; blue: seen through a blue filter **(1)**; red: longest wavelength of visible light **(1)**

3 The green object appears green because it reflects green light **(1)** and all other colours are absorbed by the object. **(1)**

4

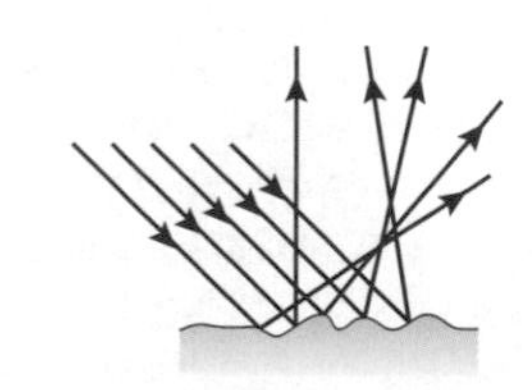

Diagram should show the incident rays striking the surfaces all parallel, with reflected rays reflected in different directions **(1)**; uneven surface shown **(1)**. Explanation should state that diffuse reflection still obeys the law of reflection because each ray of light that arrives at a surface is still reflected **(1)** according to the law of reflection (angle i = angle r) **(1)**; at the microscopic level, the surface is not smooth/the microscopic surfaces are at different angles to each other. **(1)**

35 Lenses and power

1 A converging lens bends the rays of light towards one another **(1)**; a diverging lens bends the rays of light away from each other. **(1)**

2 (a) If the lens is thicker the focal point will be closer to the lens/shorter **(1)**. If the lens is thinner the focal point will be further away from the lens/longer. **(1)**
(b) Either: the greater the power of the lens, the more it will bend light **(1)** or the greater the power of the lens, the shorter the focal length. **(1)**

3 (a) Thick lens: mid-ray continues unchanged; top and bottom rays angle down steeply **(1)**; focal point close to lens **(1)**. Thin lens: mid-ray continues unchanged; top and bottom rays angle down less steeply than for lens 1 **(1)**; focal point further from lens. **(1)**
(b) Lens 1 is more powerful. **(1)**

4 (a) Three construction rays **(1)**; focal point correctly labelled **(1)**; focal length correctly labelled. **(1)**

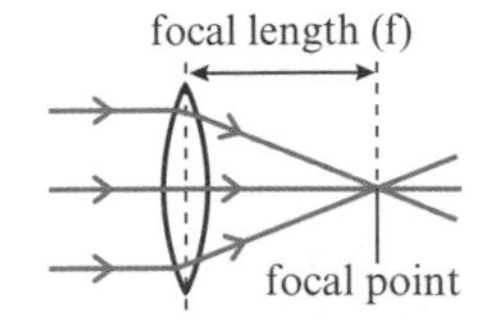

(b) Three construction rays **(1)**; divergent rays extrapolated **(1)**; focal length correctly labelled. **(1)**

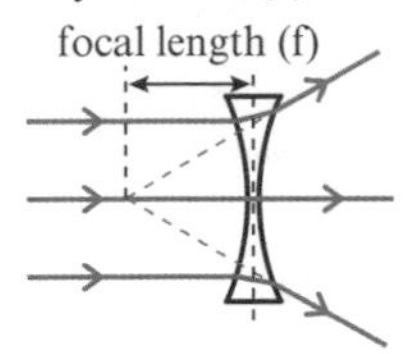

36 Real and virtual images

1 A real image is an image that can be projected on a screen **(1)** but a virtual image cannot be projected on a screen. A virtual image is produced when light **(1)** rays appear to come from a point beyond the object. **(1)**

2 principal focus **(1)**

3 (a) points added at the distance 2F on both sides of the lens **(1)**
(b) point added at the same distance as F on the right-hand side of the lens **(1)**
(c) vertical arrow drawn from where the two rays of light cross on the right-hand side up to the central line **(1)**

4 (a) larger **(1)**
(b) closer than F **(1)**
(c) at 2F **(1)**

37 Electromagnetic spectrum

1 B **(1)**

2 (a) All parts of the electromagnetic spectrum are transverse waves **(1)** and they all travel at 3×10^8 m/s/the same speed in a vacuum. **(1)**
(b) The different waves carry different amounts of energy. **(1)**

3 A: X-rays **(1)**; B: visible light **(1)**; C: microwaves **(1)**

4 $v = f \times \lambda$, so $f = v \div \lambda$ **(1)** $= 3 \times 10^8$ m/s $\div$ 240 m **(1)** $= 1.25 \times 10^6$ Hz **(1)**

38 Investigating refraction

1 (a) Place a refraction block on white paper and connect a ray box to an electricity supply **(1)**; switch on the ray box and set it at an angle to the surface of the block **(1)**; use a sharp pencil to draw around the refraction block and make dots down the centre of the rays either side of the block **(1)**; use a sharp pencil and ruler to join the 'external' rays and then draw a line across the outline of the block to join the lines **(1)**; use a protractor to draw a normal where the light ray met the block and measure the angle of incidence and angle of refraction. **(1)**
(b) When a light ray travels from air into a glass block, its direction changes **(1)** and the angle of refraction will be less than the angle of incidence. **(1)**
(c) (i) The ray of light would not change direction. **(1)**
(ii) The light would slow down **(1)** (travelling from a less dense medium to a more dense medium) and the wave fronts would be closer together/the wavelength of light would be shorter. **(1)**

2 any four from: use of electricity: if mains electricity is used there is a risk of shock – use tested apparatus/do not try to plug in/unplug in the dark **(1)**; experiments are generally done in low-level light so there is a risk of tripping – clear floor area and working space (no trailing wires) and avoid moving around too much **(1)**; if glass blocks are used there is a risk of cuts – handle with care, use Perspex/non-breakable blocks when possible **(1)**; ray boxes get hot so risk of burns – do not touch ray boxes during operation. **(1)**

3 The waves travel more slowly and wavelength becomes shorter in shallow water **(1)**; the waves change direction/bend towards the 'normal' as they move into shallower water. **(1)**

39 Wave behaviour

1 C **(1)**

2 (a) Reflection: waves bounce off a surface **(1)**; refraction: waves change speed and direction when passing from one material to another **(1)**; transmitted: electromagnetic waves are transmitted when they pass through a material **(1)**; absorbed: different electromagnetic waves are absorbed by different materials. **(1)**
(b) Any two valid examples, e.g. reflection: light on a mirror **(1)**; refraction: light through water **(1)**; transmission: radio waves passing through the atmosphere from transmitter to receiver/X-rays absorbed by the atmosphere. **(1)**

3 (a) Microwaves are shorter in wavelength **(1)** and higher in frequency **(1)** than radio waves.
(b) Microwaves sent from the ground transmitter are able to pass through the ionosphere **(1)** and are received and re-emitted by the receiver to the ground **(1)**. Radio waves sent from the ground transmitter are refracted by the ionosphere **(1)** and then reflected back to the receiver on the ground. **(1)**

4 As **oscillating** charges move up and down a radio aerial oscillating **electric** and magnetic fields move from the antenna, across space (1 mark – both words needed). When the oscillating electric **field** encounters another aerial, it causes oscillations in the receiving **electrical** circuits (1 mark – both words needed).

5 Space-based telescopes are outside the Earth's atmosphere **(1)** so they are able to detect the whole range of electromagnetic waves **(1)** that are emitted by stars and galaxies. **(1)**

40 Temperature and radiation

1 B **(1)**

2 The loaf wrapped in white paper will stay warmer for longer because the thermal energy will be reflected back inside the pack **(1)**. The black paper will emit more thermal energy to the environment. **(1)** (No marks if black paper chosen).

3 When there is a temperature difference between an object and its surroundings there will be a net transfer of thermal energy between the object and its surroundings by infrared radiation **(1)**. As the potato has a higher temperature than its surroundings **(1)**, there will be a net transfer of thermal energy by infrared radiation from the potato. **(1)**

4 Astronomers measure the intensity of each wavelength from a star's light to determine the peak (main) wavelength of light emitted from stars **(1)**. They can then relate this to a scale of colours emitted at certain temperatures and thus determine the temperature of the star. **(1)**

41 Thermal energy and surfaces

1 (a) any four from: Fill Leslie's cube with hot water at a known temperature (e.g. wait until it falls to 80 °C before taking temperature measurements) **(1)**. Measure the temperature at a distance (e.g. 10 cm) **(1)** from one of the four sides of Leslie's cube for a period of time (e.g. 5 minutes) **(1)**. Take regular readings (e.g. every 30 seconds) **(1)**. Repeat method for the other three sides. **(1)**
(b) independent variable: sides of Leslie's cube **(1)**; dependent variable: temperature **(1)**
(c) Any four from: starting temperature of the water, distance of heat sensor/thermometer from the cube, length of time, same number of readings taken, same intervals of time for each temperature reading, use the same thermometer or temperature sensor. All four correct for **2 marks**; three correct for **1 mark**; two or fewer correct for **0 marks**.

2 (a) and (b) Any one hazard and method of minimising it from: Hot water in eyes can cause damage **(1)** – always wear eye protection **(1)**. Boiling water can cause scalds **(1)** – place the kettle close to cube and fill the cube in its place **(1)**. The cube can cause burns **(1)** – do not touch the cube until the temperature reading is low **(1)**. Water and electricity can result in a shock **(1)** – when using an electrical temperature sensor, keep well away from water **(1)**. Trailing wires can be a trip hazard **(1)** – avoid trailing leads/tuck leads out of the way. **(1)**

3 (a) The bungs would minimise thermal energy transferred from the flasks through evaporation. **(1)**

(b) (i) Dull and black surfaces are the best emitters and best absorbers. **(1)**
(ii) Shiny and light surfaces are the worst emitters and worst absorbers. **(1)**

42 Dangers and uses

1 (a) A and C **(1)**
(b) B and D **(1)**
(c) A and D **(1)**

2 Some electromagnetic waves can be dangerous. Microwaves can heat **(1)** the water inside our bodies causing significant damage to cells. Infrared light waves transfer thermal **(1)** energy and can cause burns to skin. Ultraviolet waves can damage eyes **(1)** and can cause skin cancer. **1 mark** for each correct sentence.

3 X-rays and gamma rays can cause damage to DNA in cells/produce free radicals that can damage DNA **(1)**. This may lead to cell death or cause cancer. **(1)**

4 Any three from the following, X-rays are useful because they can be used to diagnose injuries without surgery **(1)** / check for medical conditions without surgery **(1)**. X-rays can be harmful to cells in the body **(1)** / can damage DNA in cells **(1)**. The use of X-rays should be controlled by carefully recording the number of X-rays delivered over time to prevent overexposure **(1)** / using ultrasound scans when risk is high (e.g. foetal scanning) to prevent over exposure to X-rays **(1)**.

43 Changes and radiation

1 D **(1)**

2 (a) Energy depends on frequency **(1)** and the higher the frequency, the greater the energy carried by the wave. **(1)**
(b) When an electron absorbs electromagnetic radiation it moves up one or more energy levels in the atom or may leave the atom entirely. **(1)**
(c) When an electron emits electromagnetic radiation it moves down one or more energy levels in the atom. **(1)**

3 The electron orbits in the atom are at specific energy levels **(1)**. An electron has to gain enough energy to jump to one or more of these levels **(1)**. If the electromagnetic radiation does not have exactly the right amount of energy the electron cannot move to higher-level orbits/too much energy will cause the electron to be ejected from the atom. **(1)**

4 Like electrons in their energy shells, protons and neutrons also occupy energy levels in the nucleus **(1)**. When energy changes occur in the nucleus, high-energy/high-frequency gamma photons are emitted **(1)** due to greater energy levels occurring in the nucleus. **(1)**

44 Extended response – Light and the electromagnetic spectrum

*Answer could include the following points.

- X-rays and gamma rays are both transverse waves.
- X-rays and gamma rays have high frequency and therefore carry high amounts of energy.
- X-rays and gamma rays cause ionisation in atoms and exposure can be dangerous/cause cells to become cancerous.
- People who work regularly with X-rays and gamma rays should limit their exposure by using shields or by leaving the room during use.
- Low-energy X-rays are transmitted by normal body tissue but are absorbed by bones and other dense materials such as metals.
- Higher-energy gamma rays mostly pass through body tissue but can be absorbed by cells.
- X-rays and gamma rays can be used to investigate/treat medical problems.
- X-rays can be used to 'see' inside containers, e.g. at border controls or at airport security/to check cracks in metals, but gamma rays are not used for this purpose.
- Gamma rays can be used to kill bacteria in food/on surgical instruments. X-rays are not used for this purpose.

45 Structure of the atom

1 (a) protons labelled in the nucleus (+ charge) **(1)**
(b) neutrons labelled in nucleus (0 charge) **(1)**
(c) electrons labelled as orbiting (– charge) **(1)**

2 (a) The number of positively charged protons in the nucleus **(1)** is equal to the number of negatively charged electrons orbiting the nucleus. **(1)**
(b) The atom will become a positively charged ion/charge of +1. **(1)**

3 (a) A molecule is two or more atoms bonded together. **(1)** (In the kinetic theory of gases, molecule also describes monatomic gases).
(b) (i) any pure liquid, e.g. water/H_2O **(1)**
(ii) any gaseous molecule, e.g. oxygen/O_2 **(1)**
(iii) any gaseous compound, e.g. carbon dioxide/CO_2 **(1)**

4 nucleus: 10^{-15} m **(1)**; atom: 10^{-10} m **(1)**; molecule: 10^{-9} m **(1)**

46 Atoms and isotopes

1 (a) the name given to particles in the nucleus **(1)**
(b) the number of protons in the nucleus **(1)**
(c) the total number of protons and neutrons in the nucleus **(1)**

2 C **(1)**

3 Isotopes will be neutral because the number of positively charged protons **(1)** still equals the number of negatively charged electrons. **(1)**

4 $^{39}_{19}K$: mass number 39 **(1)**; atomic number 19 **(1)**

5 C **(1)** (Note that the electron is considered to have 0 relative mass in calculating the atomic mass of an atom, although the relative mass of an electron is actually almost 0).

6 They both have 8 protons/they have the same proton number/atomic number **(1)**; they both have 8 electrons orbiting the nucleus **(1)**; the first has 8 neutrons while the second has 10/they have different numbers of neutrons. **(1)**

47 Atoms, electrons and ions

1 A **(1)**

2 (a) When an atom absorbs electromagnetic radiation an electron **(1)** moves to a higher energy level. **(1)**
(b) When an atom emits electromagnetic radiation an electron **(1)** moves to a lower energy level. **(1)**

3 (a) Atoms: Li, Cu **(1)**; Ions: F^-, Na^+, B^+, K^+ **(1)**
(b) Atoms are neutral and have no overall charge. **(1)** Ions have gained (–) or lost (+) an electron / have become negatively or positively charged. **(1)**

4 Any two from: An atom can lose an electron by friction (electrostatics) **(1)**. An atom can be made to lose an electron by ionising radiation (radioactivity) **(1)**. By diagram: atom with an electron being removed with force arrow labelled friction **(1)**; atom absorbing a photon with an electron being ejected. **(1)**

48 Ionising radiation

1 B **(1)**

2 alpha: very low, stopped by 10 cm of air; beta minus: low, stopped by thin aluminium; neutron: high; gamma: very high, stopped by very thick lead. All four correct for **4 marks**; three correct for **2 marks**; two correct for **1 mark**.

3 (a) no change **(1)**
(b) high-energy electron **(1)**
(c) moderately ionising **(1)**

4 (a) beta-plus (positron) **(1)**
(b) alpha particle **(1)**
(c) neutron **(1)**

5 Compared to other types of ionising radiation, the chances of collision with air particles at close range is high **(1)** because the alpha particles have a large positive charge/are massive compared to other types of radiation **(1)**. Once an alpha particle has collided with another particle it loses its energy. **(1)**

49 Background radiation

1 Radon is a radioactive element **(1)** that is produced when uranium in rocks decays. **(1)**

2 Levels can vary because of the different rocks that occur naturally in the ground **(1)**. They can also vary due to the use of different rocks such as granite in buildings. **(1)**

3 natural: any two from: air, cosmic rays, rocks in the ground, food **(1)**; manufactured: any two from: nuclear power; medical treatment; nuclear weapons **(1)**

4 (a) south-east 0.27 Bq **(1)**; south-west 0.30 Bq **(1)**
(b) south-west **(1)**

5 (a) As the uranium in rocks decays radon gas seeps out **(1)** from the soil and into homes and buildings. **(1)**
(b) When radon gas is inhaled, the alpha particles can be absorbed by the lungs **(1)** and can be ionising/dangerous in large amounts. **(1)**

50 Measuring radioactivity

1 Nuclear industry workers wear a film badge **(1)** which becomes darker when exposed to radiation **(1)**. This monitors levels of radiation to which the workers are exposed. **(1)**

2 A thin wire is connected to +400 V **(1)**. Atoms of argon are ionised **(1)**. Electrons

travel towards the thin wire **(1)**. The amount of radiation detected is shown by the rate meter. **(1)**

3 The student is correct. When radiation is more ionising, it is more likely to create ions **(1)** so more highly ionising radiation is more likely to ionise the argon in the tube **(1)**, which means that it is more likely to cause a current and be recorded on the rate meter. **(1)**

4 Aluminium absorbs beta particles **(1)** and lead absorbs gamma rays **(1)** so this enables the type of radiation to which the wearer has been exposed to be identified. **(1)**

51 Models of the atom

1 The plum pudding model showed the atom as a 'solid', positively charged **(1)** particle containing a distribution of negatively charged electrons **(1)** while the Rutherford model showed the atom as having a tiny, dense, positively charged nucleus **(1)** surrounded by orbiting negatively charged electrons. **(1)**

2 Rutherford fired positively charged alpha particles at atoms of gold foil; most went through showing that there were large spaces in the atom. **(1)** Some were repelled or deflected **(1)** showing that the nucleus was positively charged. **(1)**

3 (a) A **(1)**
(b) The Bohr model showed that electrons **(1)** orbit the atom at different energy levels **(1)** and those electrons have to acquire precise amounts of energy to move up to higher levels **(1)**. The model was an improvement because it was able to explain emission and absorption spectra which enables the atom to be stable. **(1)**

4 (a) Electrons can absorb specific amounts of energy from photons **(1)**. These transfer energy to the electrons which become excited/have more energy so they move to a higher energy level. **(1)**
(b) A photon, of equivalent energy to the energy lost from the 'excited' electron, is emitted from the atom. **(1)**

52 Beta decay

1 Beta-minus decay is when a neutron (n) changes to a proton releasing a high-energy electron (e^-). **(1)** Beta-plus decay is when a proton (p) changes to a neutron (n) releasing a high-energy positron (e^+). **(1)**

2 (a) 7 **(1)**
(b) 12 **(1)**

3 (a) In beta-minus decay, a neutron decays into a proton **(1)** and a high-energy electron (beta-minus particle) is emitted. **(1)**
(b) In beta-plus decay, a proton decays into a neutron **(1)** and a positron (beta-plus particle) is emitted. **(1)**

4 In archaeology, beta decay is used to date objects using radioactive carbon dating. **(1)** In medicine, beta decay is used for producing images in positron (PET) scanning. **(1)**

53 Radioactive decay

1 B and C **(1)**

2 beta-negative, charge (−1) **(1)**; beta-positive, charge (+1) **(1)**

3 The total mass number before a reaction/decay must be the same as the total mass number after a reaction/decay **(1)** so in any nuclear decay the total mass number on either side of the decay equation must balance. **(1)**

4 In neutron decay, a neutron **(1)** is emitted and a new isotope of the element is formed. **(1)**

5 5 B **(1)**

6 (a) (i) Add 208 to Po **(1)**; alpha **(1)**
(ii) Add 86 to Rn **(1)**; alpha **(1)**
(iii) Add 42 to Ca **(1)**; beta-minus **(1)**
(iv) Add 9 to Be **(1)**; neutron **(1)**
(b) Nucleons often rearrange themselves **(1)** following alpha or beta decay. This causes energy to be emitted as a gamma photon / wave. **(1)**

54 Half-life

1 (a) 8 million atoms **(1)**
(b) 9.3 ÷ 3.1 = 3 half-lives **(1)**, 1 half-life – 8 million; 2 half-lives – 4 million; 3 half-lives – 2 million atoms **(1)**

2 The activity is 400 Bq at 1.5 min **(1)** (between 1.3 and 1.7 is allowed), half this activity is 200 Bq, which is at 6.5 min **(1)** (between 6.3 and 6.7 is allowed) so the half-life is 6.5 min – 1.5 min = 5 min **(1)** (Answers between 4.7 and 5.3 min are allowed). (If you used other points on the graph and found an answer of around 5 min you would get full marks. For this question your working can just be pairs of lines drawn on the graph).

3 The prediction is based on the half-life of caesium **(1)**. You would expect that from 1986 to 2016, radioactivity would have fallen to half this level **(1)**. The level of radioactivity does not fall as rapidly because of background radiation/radioactive materials in the soil. **(1)** (Marks are awarded for discussion of the source of the prediction, i.e. the half-life of caesium, and recognising that other substances will release radiation in addition to the caesium).

55 Uses of radiation

1 (a) Beta **(1)** radiation is used because alpha particles would not pass through and gamma rays would pass too easily. **(1)**
(b) (i) The paper has become too thick. **(1)**
(ii) The pressure on the rollers would be increased to make the paper thinner. **(1)**

2 They have high frequency/they carry large amounts of energy. **(1)**

3 (a) Americium-241 is an alpha emitter; alpha particles cannot pass through a metal casing within the plastic casing of the smoke alarm **(1)**. They ionise air particles and in doing so lose their energy through about 10 cm of air. **(1)**
(b) The alpha particles ionise/charge the air (molecules) **(1)**. The charged air molecules are attracted to the plates, completing the circuit. **(1)**
(c) The smoke particles absorb the alpha particles **(1)** so the current decreases and the siren sounds. **(1)**

4 Plastic instruments cannot be heated to sterilise them **(1)** so gamma rays can be used to kill bacteria/microbes. **(1)**

56 Dangers of radiation

1 any two from: hospital **(1)**; dental surgery **(1)**; radiography/X-ray department **(1)**; nuclear power plant **(1)**

2 (a) Ionising means that electrons are removed from atoms. **(1)**
(b) Ions in the body can cause damage to cell tissue **(1)**, which can lead to DNA mutations/cancer. **(1)**

3 (a) Employers can limit the time of exposure **(1)**; workers can wear protective clothing/wear a lead apron **(1)**; increase distance from the source. **(1)**
(b) The amount of energy/dose of radiation that a person has been exposed to is monitored by wearing a film badge **(1)**. This is checked each day. **(1)**

4 The greater the half-life of an ionising source **(1)**, the longer it will remain dangerous. **(1)**

5 X-rays carry enough energy **(1)** to ionise atoms by removing electrons. **(1)**

57 Contamination and irradiation

1 (a) Radium was useful because it was luminous, allowing watches to be used in the dark. **(1)**
(b) Before 1920 the effects of radium were not known/recognised **(1)** so it was thought that it was safe to use **(1)**. It was banned from use once the dangers were known. **(1)**

2 external contamination: radioactive particles come into contact with skin, hair or clothing; internal contamination: a radioactive source is eaten, drunk or inhaled; irradiation: a person becomes exposed to an external source of ionising radiation. All correct for **2 marks**, 2 correct for **1 mark**.

3 (a) Any suitable example, e.g. contaminated soil may get on to hands. **(1)**
(b) Any suitable example, e.g. contaminated dust or radon gas may be inhaled. **(1)**

4 Internal contamination means that the alpha particles come into contact with the body through inhalation or ingestion **(1)** where they are likely to cause internal tissue damage **(1)**. Alpha particles that are irradiated are less likely to cause damage because they have to travel through air **(1)** and are therefore less likely to ionise body cells **(1)** (at distances of over 10 cm).

58 Medical uses

1 (a) C **(1)**
(b) Several beams of gamma rays are fired, from different positions, towards the location of the cancer tumour **(1)**. The cancer tumour receives the full energy and the surrounding healthy tissue receives less. **(1)**

2 A: 2 **(1)**; B: 4 **(1)**; C: 1 **(1)**; D: 5 **(1)**; E: 3 **(1)**

3 (a) tracer **(1)**
(b) The radioactive tracers decay very quickly. **(1)**

4 Both internal and external approaches use ionising radiation **(1)** which can be used to specifically target the cancer tumours **(1)**. Internal treatment involves putting the radioactive source into the body by injection/ingestion **(1)**; external treatment involves firing gamma rays from a gamma source. **(1)**

59 Nuclear power

1 It does not produce acid rain/smoke/ash during operation. **(1)** Less nuclear fuel is needed compared to fossil fuels/longer future availability. **(1)** There is no emission of CO_2 to the environment during operation. **(1)**

2 Nuclear fusion: small nuclei, such as hydrogen, are fused together under huge pressure and temperature, releasing enormous amounts of light and heat **(1)**; this occurs in stars like the Sun. **(1)** Nuclear fission: large nuclei are split by slow-moving neutrons **(1)**; this happens in a nuclear reactor using uranium, releasing enormous amounts of thermal energy. **(1)** Natural radioactive decay also releases energy but radioactive materials are put under controlled conditions to release energy for power generation through fission.

3 Any three from: It is difficult to store and contain the radioactive nuclear waste safely **(1)**. It is expensive to store nuclear waste safely **(1)**. Nuclear waste must be stored for thousands of years **(1)**. An accident would involve containment/treatment over a large area. **(1)**

4 It has been claimed that global warming can be increased by adding more carbon dioxide to the atmosphere during the construction processes of power plants **(1)** and through the production of fuel rods. The nuclear industry can claim that no carbon dioxide is produced when the power station is operating **(1)** so generating electricity does not affect climate change/global warming in the long term/as much as some other resources. **(1)**

60 Nuclear fission

1 left nucleus: uranium-235 **(1)**; two new nuclei: daughter nuclei **(1)**; small single particles: neutrons **(1)**

2 Uranium-238 has three more **(1)** neutrons **(1)** than uranium-235.

3 (a) moderator: slows down the neutrons so they can be absorbed by uranium-235 **(1)**
(b) control rods: absorb excess neutrons **(1)**

4 A chain reaction occurs as neutrons are released from a fission reaction and are absorbed by more uranium nuclei **(1)**. As more than one neutron is released per fission reaction **(1)**, this leads to more and more fission reactions occurring, which could lead to an explosion **(1)**. Control rods are used to absorb excess neutrons to keep the chain reaction at a steady rate. **(1)**

5 If neutrons are too fast they are not absorbed by the uranium-235 nuclei **(1)** and fission does not occur. **(1)**

61 Nuclear power stations

1 (a) top left: reactor **(1)**; top right: turbine **(1)**; bottom: generator **(1)**
(b) (i) reactor: nuclear store **(1)**;
(ii) turbine: kinetic store **(1)**
(c) electricity generation **(1)**

2 D **(1)**

3 Nuclear fission starts with a slow-moving neutron being absorbed by a uranium-235 nucleus **(1)**; the uranium-235 nucleus splits to produce two lighter elements **(1)** and (at least) two more neutrons. **(1)** In this process thermal energy is also released **(1)**. (Marks may be awarded for each point shown in a diagram).

4 Using nuclear fission as an energy resource produces no polluting products of combustion/gas emissions **(1)**; it is a concentrated energy resource/a large amount of energy is produced from a small amount of fuel **(1)**; highly radioactive products need safe containment/disposal **(1)**; there is a long-term need for storage of radioactive products. **(1)**

62 Nuclear fusion

1 B **(1)**

2 (a) Other scientists have been unable to repeat the experiments. **(1)**
(b) Fusion needs high pressures and temperatures **(1)**. At present this requires more energy than is released by fusion. **(1)**

3 (a) stars **(1)**
(b) high pressure **(1)** and high temperature **(1)**
(c) The electrostatic repulsion between protons needs to be overcome for fusion to occur. **(1)**

4 When fusion occurs the mass of the products is slightly less than the mass of the reactants **(1)** so this mass difference **(1)** is released as energy in the form of light and thermal energy.

63 Extended response – Radioactivity

*Answer could include the following points.

- Nuclear fusion is the creation of larger nuclei, resulting in a loss of mass from smaller nuclei, accompanied by a release of energy.
- A large amount of energy is produced from a small amount of fusion fuel.
- Fusion occurs in stars, which have high temperature and pressure, where hydrogen nuclei are fused to become helium nuclei.
- Nuclear fusion does not happen at low temperatures and pressures due to electrostatic repulsion of protons.
- Nuclear fusion requires huge amounts of energy to overcome the electrostatic repulsion of protons.
- Due to the huge amounts of energy required, it is currently not economic to use fusion to generate electricity.
- It is practically very difficult to create high-enough temperatures and pressures for fusion to occur on a large scale.
- There would still be problems with dealing with radioactive waste material although this would be much less than for fission.

64 The Solar System

1 (a) Venus **(1)**; Mars **(1)**; Saturn **(1)**; Uranus **(1)**
(b) (i) dwarf planet **(1)**
(ii) More powerful observational telescopes **(1)** have found other objects close to Pluto's orbit. **(1)**

2 Ptolemy's geocentric model **(1)** explained the Solar System as the Sun and the Moon orbiting the Earth **(1)**. Copernicus' heliocentric model **(1)** explained the Solar System as the Moon orbiting the Earth and the Earth orbiting the Sun. **(1)**

3 (a) C **(1)**
(b) Comets have highly elliptical paths whereas the others have circular or nearly circular orbits. **(1)**

4 (a) The term orbit means to travel in a circular or elliptical path around a central body. **(1)**
(b) In a circular orbit the orbiting body travels in a regular circular path **(1)** but in an elliptical orbit the body travels in an elongated oval path. **(1)**
(c) moons or natural satellites **(1)**

65 Satellites and orbits

1 C **(1)**

2 (a) Any one of: GPS (global positioning satellite) systems, telecommunications satellites. **(1)**
(b) Any one of: weather monitoring, military, spying or Earth-observation purposes. **(1)**

3 (a) distance from the central mass **(1)**
(b) (i) If the satellite's orbital speed decreased, the Earth's gravitational field would cause the satellite to fall towards the Earth **(1)**. If the speed increased, the satellite would move to a higher orbit. **(1)**
(ii) If the gravitational field strength of the Earth decreased, the velocity of the satellite would move it to a higher orbit **(1)**. If the gravitational field strength of the Earth increased, the satellite would be attracted closer to Earth and fall out of the orbit. **(1)**

4 (a) The increase in gravitational field strength as the comet approaches the Sun causes it to accelerate. **(1)**
(b) As the comet approaches the Sun, the gravitational field strength causes the comet to come close to the Sun at high speed **(1)**. As the comet moves away from the Sun, the gravitational field strength decreases **(1)** and so the comet decelerates as it moves away from the Sun **(1)**. At the furthest point from the Sun, the speed and velocity are at a minimum **(1)**. The speed starts to increase again as the comet moves back towards the Sun. **(1)**

66 Theories about the Universe

1 (a) All four correct for **3 marks**; three correct for **2 marks**; two correct for **1 mark**. Big Bang theory – A; Steady State theory – B, C, D
(b) the evidence of red shift leads to the conclusion that the emitters (stars/galaxies) are moving away from us **(1)**

2 CMB radiation is detected from everywhere, **(1)** supporting the theory of an original point source that 'exploded' outwards (the Big Bang theory). **(1)**

3 (a) When light from a distant galaxy is compared with the light observed from the Sun the (absorption) spectra/lines in the spectrum **(1)** are shifted/moved **(1)** towards the red end. **(1)**
(b) A galaxy with a red-shifted spectrum indicates that the galaxy is moving away from the observer **(1)**. The further the black lines are shifted, the faster it is moving away **(1)**. As most galaxies are red-shifted, this would suggest the universe is expanding. **(1)**

67 Doppler effect and red-shift

1 D (1)

2 At point A the siren is approaching so the pitch is higher because the waves are closer together/wavelength decreases (1). At point B the siren is moving away so the pitch is lower because the waves are further apart/ wavelength increases. (1) (Each answer must include a change in pitch and the explanation referring to wavelength or frequency).

3 (a) The light from comet A has a longer wavelength than that from comet B (1). The light from comet A has a lower frequency than the light from comet B. (1)
(b) Comet A is moving away from the Earth (1) and comet B is moving towards the Earth. (1)

4 (a) Hubble found that galaxies further away were red-shifted the most. (1)
(b) When the galaxies were plotted on a graph of speed against distance from Earth (1), Hubble found that the galaxies furthest away were travelling the fastest (1), which suggested that the Universe was expanding. (1)

68 Life cycle of stars

1 (nuclear) fusion (1)

2 When all nuclear fusion stops/the elements that undergo fusion are used up (1) the star collapses under gravity (1) to become a white dwarf. The reduction in fusion reduces the balancing outwards force of thermal expansion (1) so the force of gravity is greater, causing the collapse. (1)

3 T (1); F (1); F (1); T (1); T (1) (Five correct – **three marks**, four correct – **2 marks**, three correct – **1 mark**).

4 The mass of the star and *g* (due to the force of gravity) results in an inwards force (1). This is balanced by the production of huge amounts of thermal energy (1) causing thermal expansion (an outward force). (1)

69 Observing the Universe

1 (a) We obtain clearer images. (1)
(b) Early astronomers used their eyes to observe visible light from stars (1) and later they invented the early telescopes. (1)
(c) We can gather more data. (1)

2 Any two of: Telescopes in space are beyond the atmosphere and so are able to detect the wavelengths of the electromagnetic spectrum that are absorbed by the atmosphere (1). X-rays cannot pass through the atmosphere so X-ray telescopes are used in space (1). A space telescope can detect gamma and ultra violet, some visible light, some infrared, some microwaves and some radio waves. (1)

3 An array of telescopes can be made to work together to produce a much larger detector. (1)

4 The atmosphere is thinner at high altitudes (1) so more data can be collected by the telescope/less of the electromagnetic spectrum is absorbed by the atmosphere. (1)

5 These wavelengths are absorbed by the Earth's atmosphere (1) so this data could not have been collected by Earth-based telescopes. (1)

70 Extended response – Astronomy

*Answer could include the following points.

- The planets in the Solar System are in stable orbits and are each travelling at the right speed based on their distance from the Sun.
- The planets tend to move in circular or near-circular orbits around the Sun.
- The moons of planets tend to move in circular or near-circular orbits around their parent planet.
- If the planets' orbital speeds increased, the gravitational force from the Sun would not keep them in orbit and they would fly out of orbit.
- If the planets' orbital speeds decreased, the Sun's gravitational field would cause them to fall towards it.
- Comets travel in highly elliptical orbits around the Sun.
- The gravitational force of the Sun acting on the comet gets weaker as it gets further away from the Sun.
- The comet will travel slower when it is far from the Sun and faster when it is closer to the Sun.
- Artificial (man-made) satellites orbiting the Earth, and planets orbiting the Sun, tend to move in circular or near-circular orbits.
- Satellites in geostationary orbits move so that they are in a fixed position over the Earth, e.g. GPS (global positioning satellite) systems.
- Satellites in low polar orbits move over the poles of the Earth, e.g. weather monitoring, military, spying or Earth-observation purposes.

71 Work, energy and power

1 D (1)

2 (a) gravitational potential energy store (1)
(b) thermal energy store (1)
(c) chemical energy store (1)

3 energy transferred = 15 000 J, time taken = 20 s; $P = E \div t$ = 15 000 J ÷ 20 s (1) = 750 (1) W (1)

4 work done = $F \times d$ = 600 N × (20 × 0.08 m) (1) = 960 (1) J (1)

5 $P = E \div t$ so $t = E \div P$ (1) = 360 000 J ÷ 200 s (1) = 1800 s (1) (or 30 minutes)

72 Extended response – Energy and forces

*Answer could include the following points.

- This is described as a mechanical process because the turbine is moved by the kinetic energy of the moving air.
- Mechanical energy is the sum of potential and kinetic energy.
- The mechanical energy of the moving air gives the air particles the ability to apply a force and cause a displacement of the blades.
- Mechanical processes become wasteful when they cause a rise in temperature so dissipating energy to the thermal store of the environment.
- Rise in temperature is caused by friction between moving objects/materials.
- It is important to keep friction as low as possible to minimise wasted energy.
- By reducing wasted energy the wind turbines can be made more efficient.
- Higher efficiency will mean more electricity is generated.

73 Interacting forces

1 (a) gravitational (1), magnetic (1), electrostatic (1)
(b) Gravitational fields are different because they only attract (1) whereas magnetic and electrostatic attract and repel. (1)

2 A, (1) C (1)

3 Weight is a vector because it has a direction (downwards) (1). Normal contact force is a vector because it has a direction (upwards/ opposite to weight). (1)

4 (a) The horizontal contact forces are pull (by the student on the bag) and friction (of the bag against the floor) (1). The forward pull force on the bag is equal to the opposing frictional force (1) (so the bag moves at constant velocity). (1)
(b) weight and normal contact/reaction force (1)

74 Free-body force diagrams

1 B (1)

2 (a) arrow above bird pointing upward (1) arrow below bird pointing downward (1), (both arrows must be the same length).
(b) Reaction force of branch upwards = 20 (1) N (1); weight downwards = 20 (1) N (1)

3 **1 mark** for each arrow. Longest arrow to left indicating resultant is forwards.

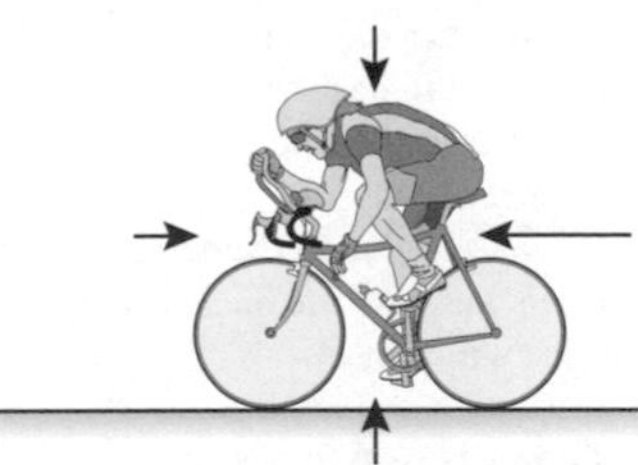

4 (a) 3.6 cm (1)
(b) 7.2 N (1)

75 Resultant forces

1 (a) A: 9.5 N (1); B: 2 N (1); C: 4.5 N (1); D: 12.75 N (1)
(b) A: up (1); B: up (1); C: to the right (1); D: to the left (1)

2 B (1)

3 (a) diagonal arrow: 2.5 cm (1)
(b) 50 N (1)

4 scale correct, e.g. vertical 2 cm (6 N) and horizontal 5 cm (15 N) (1); hypotenuse = 5.4 cm (1) represents 16.2 N (1)

76 Moments

1 B (1)

2 When an object is balanced the clockwise moment (1) is equal to the anticlockwise moment. (1)

3 $M = F \times d$ = 25 N × 0.28 m (1) = 7 (1) Nm (1)

4 (a) No (1); moment for Alex = 300 N × 0.8 m = 240 Nm (1); moment for Priya = 250 N × 1.2 m = 300 Nm. (1)
(b) Moment for Alex must change to equal 300 Nm so distance must change to 300 Nm ÷ 300 N = 1 m. (1)
(c) Original moment for Alex = 240 Nm so new moment for Priya must equal this (1); 240 Nm ÷ 250 N = 0.96 m (1), so Priya must move to 0.96 m from the pivot.

77 Levers and gears

1 B **(1)**

2 class 1 levers: hammer, scissors **(1)**; class 2 levers: bottle opener, nutcracker **(1)**; class 3 levers: broom, tongs. **(1)** (Both correct answers needed for each mark).

3 For a high gear, the driver gear has a greater diameter than the driven gear **(1)** and the output is low for a given input force. **(1)** For a low gear, the driver gear has a similar diameter to the driven gear **(1)** and the output is high for a given input force. **(1)**

4 The cyclist would change to a low gear when moving from a horizontal road to a hill because a high output is needed **(1)** for a given force to move the bicycle up the hill against gravity. **(1)**

78 Extended response – Forces and their effects

*Answer could include the following points.

- When the drone takes off the downwards force of the rotor blade creates a reaction force (Newton's third law).
- The reaction force is greater than the downwards force of the weight of the drone and it moves upwards.
- The vertical resultant force continues during the flight of the drone.
- The height can be adjusted by increasing or decreasing the upwards reaction force (due to the rotor blades).
- To move to a location, an additional resultant horizontal force is required.
- As the drone flies horizontally, the thrust of the drone must be greater than the air resistance acting in the opposite direction.
- When these two are balanced or zero the drone will hover.

79 Circuit symbols

1 When there is an electric current in a resistor, there is an energy transfer which heats the resistor. **(1)**

2 (a) C, **(1)** D **(1)**
(b) (i) The thermistor responds by changing resistance with changes in temperature. **(1)**
(ii) The LDR responds by changing resistance with changes in light intensity. **(1)**

3

Component	Symbol	Purpose
ammeter	A	measures electric current **(1)**
fixed resistor		provides a fixed resistance to the flow of current **(1)**
diode		allows the current to flow one way only **(1)**
switch	or	allows the current to be switched on or off **(1)**

4 Diagram showing series circuit diagram with battery **(1)**; ammeter in series **(1)** and switch in series **(1)**; thermistor in series **(1)**; motor in series **(1)** voltmeter connected across the motor in parallel. **(1)**

80 Series and parallel circuits

1 (a) series: $A_2 = 3$ A **(1)**; $A_3 = 3$ A **(1)**; parallel: $A_2 = 1$ A **(1)**; $A_3 = 1$ A **(1)**; $A_4 = 1$ A **(1)**
(b) In a series circuit the current is the same throughout the circuit **(1)**. In a parallel circuit the current splits up in each branch. **(1)**

2 (a) series: $V_2 = 3$ V **(1)**; $V_3 = 3$ V **(1)**; $V_4 = 3$ V **(1)**; parallel: $V_2 = 9$ V **(1)**; $V_3 = 9$ V **(1)**; $V_4 = 9$ V **(1)**
(b) In a series circuit the potential difference is shared/splits up across the components in the circuit **(1)**. In a parallel circuit the potential difference across each branch is the same as the supply potential difference. **(1)**

3 (a) In a parallel circuit, each component is supplied with sufficient potential difference to work properly **(1)**. If a fault develops, other parts of the circuit will still work. **(1)**
(b) The lamps would share the potential difference in the circuit so each component would not operate at full capacity **(1)**. If a fault developed, the whole supply would be cut as there would be no other route for the current to take. **(1)**

81 Current and charge

1 (a) An electric current is the rate **(1)** of flow of charge (electrons in a metal). **(1)**
(b) $Q = I \times t = 4\text{ A} \times 8\text{ s}$ **(1)** = 32 **(1)** coulombs/C **(1)**

2 (a) (i) $A_1 = 0.3$ A **(1)**
(ii) $A_3 = 3$ A **(1)**
(b) add another cell/increase the energy supplied **(1)**
(c) The electrons move around the circuit in one continuous path **(1)** so the current leaving the cell is the same as the current returning to it. **(1)**

3 (a) Any series circuit diagram with a component (e.g. lamp) **(1)** and an ammeter. **(1)**
(b) stopwatch/timer **(1)**

82 Energy and charge

1 Current is the charge flowing per unit time. **(1)** Potential difference is the energy transferred per unit of charge. **(1)**

2 $E = Q \times V$ **(1)** = 30 C × 9 V **(1)** = 270 J **(1)**

3 $E = Q \times V$ so $Q = E \div V$ **(1)** = 125 J ÷ 5 V **(1)** = 25 C **(1)**

4 $Q = E \div V$ = 600 J ÷ 20 V **(1)** = 30 C **(1)**; $Q = I \times t$, so $t = Q \div I$ = 30 C ÷ 0.15 A **(1)** = 200 s (or 3 min 20 s) **(1)**

83 Ohm's law

1 D **(1)**

2 Ohm's law means that the rate of flow of electrons (the current) flowing through the resistor **(1)** is directly proportional to the potential difference across the resistor. **(1)**

3 (a) $R = V \div I$ = 12 V ÷ 0.20 A **(1)** = 60 Ω **(1)**
(b) R = 22 V ÷ 0.40 A **(1)** = 55 Ω **(1)**
(c) R = 9 V ÷ 0.03 A **(1)** = 300 Ω **(1)**
(d) resistor in part (c) **(1)**

4 (a) Line A: straight line through origin **(1)**; line B: straight line through origin, different gradient. **(1)**
(b) the line with the lower gradient **(1)**

84 Resistors

1 A **(1)**

2 (a) 20 + 30 + 150 **(1)** = 200 Ω **(1)**
(b) (i) The sum of the potential differences across the resistors connected in series must equal the potential difference across the battery. **(1)**
(ii) R_T = 200 Ω, I_T = 0.03 A; $V = I \times R$ = 0.03 A × 200 Ω **(1)** = 6 V **(1)**; two identical cells so each cell supplies 3 V **(1)**

3 In a parallel circuit there are two paths for the current to take **(1)**; more current takes the path of least resistance so the value of the current is higher. **(1)** (Accept the converse.)

85 I–V graphs

1 (a) C **(1)**
(b) As the potential difference increases the current increases **(1)** in a linear/proportional relationship. **(1)**
(c) As the potential difference increases the current increases **(1)** but the gradient of the line gets less steep/shallower or the increase in current becomes smaller as potential difference continues to increase. **(1)**

2 (a) Fixed resistor: same as graph A in Q1 **(1)**; filament lamp: same as graph B in Q1. **(1)**
(b) The graphs are a different shape from each other because the fixed resistor (at constant temperature) is ohmic/obeys Ohm's law so the current and potential difference have a proportional relationship **(1)**. The filament lamp does not obey Ohm's law, as temperature increases, so the relationship between current and potential difference is not proportional. **(1)**

3 Data can be collected using an ammeter to measure current **(1)** and a voltmeter to measure potential difference **(1)**. A variable resistor **(1)** should be included which will allow different values of current to be obtained **(1)**. Resistance can then be calculated from Ohm's law. **(1)**

86 Electrical circuits

1 (a) Two resistors in same loop **(1)**; at least one ammeter shown and connected in series **(1)** at least one voltmeter shown and connected in parallel across a resistor or cell/battery. **(1)**

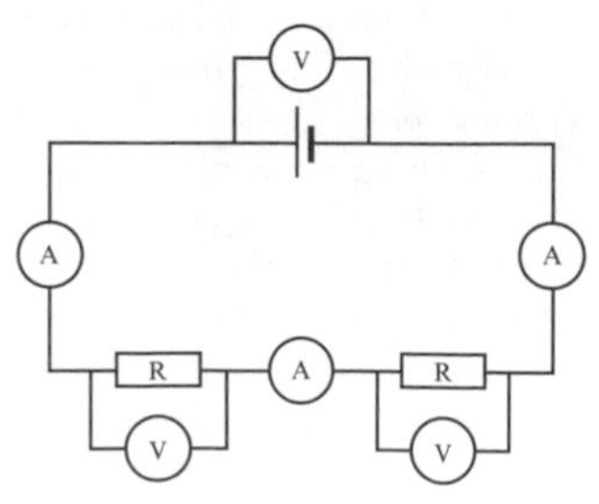

(b) two resistors in separate loops **(1)**; at least one ammeter shown and connected in main circuit or in a loop connected in series **(1)**; at least one voltmeter shown and connected in parallel across a resistor or cell/battery **(1)**

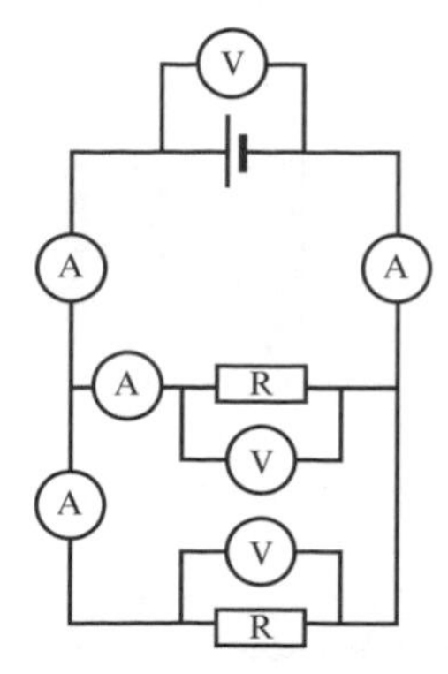

(c) Current is the same at any point in a series circuit **(1)** but will split up at a junction in a parallel circuit **(1)**. The sum of the potential difference across components in a series circuit equals the potential difference of the cell **(1)**. The sum of the potential difference across components in each loop in a parallel circuit equals the potential difference of the cell. **(1)**

2 (a) Connect a cell, a filament lamp, a variable resistor and an ammeter in a series circuit **(1)**. Connect the voltmeter across the filament lamp in parallel **(1)**. Adjust the variable resistor setting to obtain a number of readings for current and potential difference. **(1)**

(b) Ohm's law: resistance = potential difference ÷ current ($R = V \div I$) **(1)**

(c) Plotting a graph of I against V shows how the current through it varies with the potential difference across it. The gradient of the I–V graph is equal to $1/R$ **(1)**, so inverting the value of the gradient gives you the resistance, R. **(1)**

3 One from: Resistors can become hot and cause burns **(1)** or fire. **(1)**

87 The LDR and the thermistor

1

Light-dependent resistor (LDR)	Thermistor
(1)	(1)

2 D **(1)**

3 (a) The resistance goes down (more current flows) as the light becomes more intense (brighter). **(1)**

(b) The resistance goes down (more current flows) as the temperature goes up. **(1)**

4 The lamp lights up when the temperature is high **(1)** because the current through the lamp and the thermistor will be high when the resistance of the thermistor falls. **(1)**

5 When the level of light increases, the resistance decreases **(1)** and the current increases. **(1)**

88 Current heating effect

1 C **(1)**

2 When a conductor is connected to a potential difference the free electrons **(1)** move through the lattice of metal ions **(1)**. As they do so, collisions **(1)** occur where the kinetic energy is transferred into thermal energy **(1)**, causing the heating effect.

3 Any three suitable examples, e.g. electric fire **(1)**, hairdryer **(1)**, kettle **(1)**, iron **(1)**, toaster. **(1)**

4 All the appliances draw a certain amount of current **(1)**. Those that have heating elements draw more **(1)**. Even though the plugs are earthed, too much current being drawn **(1)** could cause a heating effect/fire in the multi-socket. **(1)**

5 Domestic filament lamps that were only about 5% efficient would have transferred about 95% **(1)** of the electrical energy as 'wasted' thermal energy **(1)**. The lamps were withdrawn and replaced with lamps that transferred more useful energy as light. **(1)**

89 Energy and power

1 (a) Using the equation for power $P = I \times V$ = 5 A × 230 V **(1)** = 1150 **(1)** W

(b) $E = I \times V \times t$ = 0.2 A × 4 V × 30 s **(1)** = 24 **(1)** J **(1)**

2 (a) $P = I \times V$ so $I = P \div V$ **(1)** = 3 W ÷ 6 V **(1)** = 0.5 A **(1)**

(b) $E = I \times V \times t$ = 0.5 A × 6 V × 300 s **(1)** = 900 **(1)** J **(1)** (or $E = P \times t$ = 3 W × 300 s **(1)** = 900 **(1)** J **(1)**)

(c) $P = I^2 \times R$ **(1)** = $(0.5\ \text{A})^2 \times 240\ \Omega$ **(1)** = 60 W **(1)**

90 A.c. and d.c. circuits

1 (a) An alternating current is an electric current that changes direction regularly **(1)** and its potential difference is constantly changing. **(1)**

(b) A direct current is an electric current in which all the electrons flow in the same direction **(1)** and its potential difference has a constant value. **(1)**

2 (a) $P = E \div t$ so $E = P \times t$ = 2000 × (15 × 60) s **(1)** = 1 800 000 J **(1)**

(b) E = 1500 W × 25 s **(1)** = 37 500 J **(1)**

(c) E = 10 × (6 × 60 × 60) s **(1)** = 216 000 J **(1)**

3 (a) The current is a direct current **(1)** because the electrons all flow in the same direction. **(1)**

(b) There should be one horizontal line anywhere on the screen. **(1)**

91 Mains electricity and the plug

1 (a) earth wire (green and yellow) **(1)**; live wire (brown) **(1)**; neutral wire (blue) **(1)**; fuse **(1)**

(b) The fuse is connected to the live wire **(1)** because it carries the current into the appliance. **(1)**

2 brown: Electrical current enters the appliance at 230 V **(1)**. blue: Electrical current leaves the appliance at 0 V through this wire **(1)**. green/yellow: This is a safety feature connected to the metal casing of the appliance. **(1)**

3 When a large current enters the live wire **(1)** this produces thermal energy **(1)** which melts the wire in the fuse **(1)**. The circuit is then broken. **(1)**

4 (a) When a current is too high **(1)** a strong magnetic field is generated which opens a switch (held back by a spring) **(1)**. This 'breaks' the circuit, **(1)** making it safe.

(b) The earth wire is connected to the metal casing **(1)**. If the live wire becomes loose and touches anything metallic **(1)** the current passes through the earth wire. **(1)**

92 Extended response – Electricity and circuits

*Answer could include the following points.

- The thermistor can be connected in series with an ammeter to measure current with a voltmeter connected in parallel across it to measure potential difference.
- Ohm's law can be referred to in calculating the resistance.
- When the temperature is low the resistance of the thermistor will be high, allowing only a small current to flow.
- When the temperature is high the resistance of the thermistor will be low, allowing a larger current to flow.
- The light-dependent resistor can be connected in series with an ammeter to measure current with a voltmeter connected in parallel across it to measure potential difference.
- When light levels are low (dark) the resistance of the light-dependent resistor will be high, allowing only a small current to flow.
- When light levels are high (bright) the resistance of the light-dependent resistor will be low, allowing a larger current to flow.
- Thermistors can be used in fire alarms as a temperature sensor to switch on an alarm.
- Light-dependent resistors can be used in security systems as a light sensor to switch on a light.

93 Static electricity

1 The student transfers a charge onto the balloon by transferring electrons from the jumper to the balloon **(1)**. The negative charges on the balloon repel the electrons in the wall **(1)** inducing a positive charge on the wall **(1)**, which attracts the negatively charged balloon. **(1)**

2 B **(1)**

3 (a) Friction transfers electrons to another material. **(1)**

(b) Friction transfers electrons from another material. **(1)**

4 Insulators do not allow electrons to flow. **(1)** Instead the electrons either collect on the insulator (building up a charge) or are knocked off (leaving a positive charge) **(1)**. Conductors allow electrons to flow **(1)**. Mutually repelling electrons will then flow away and this dissipates any charge build-up. **(1)**

5 (a) Electrons have been transferred by friction leaving opposite, attracting charges **(1)** between the cloth and the rod. **(1)**

(b) The student could suspend the second rod so it rotates freely **(1)**. By moving the first rod close, the second rod should swing away showing repulsion **(1)** because like charges repel. **(1)**

94 Electrostatic phenomena

1 All five correct for **4 marks**, three correct for **3 marks**, three correct for **2 marks**, two correct for **1 mark**. E, C, A, B, D

2 When the student walked along the carpet he gained electrons by friction because his trainers are insulators **(1)**. His trainers insulate him, so the charge builds up on his body **(1)**. When he touched the door handle the static charge discharges to the metal and causes a spark/shock **(1)**.

3 (a) Convection currents in the thunderclouds cause air, water and ice particles to move against each other **(1)**. The friction removes electrons **(1)** which cannot escape to earth as they are insulated by the air **(1)**. When many electrons build up the charge becomes big enough to arc/jump to earth in a lightning strike. **(1)**
(b) The negative electrons in the charged thundercloud are attracted to the earth **(1)**, so they take the nearest route (the tallest building) **(1)**. The conducting metal of the lightning rod provides the path of lowest resistance to earth. **(1)** When tall buildings are struck by lightning the electrons are conducted down the lightning rod. **(1)**

95 Electrostatics uses and dangers

1 (a) As the aircraft is refuelled, the pipe becomes charged by collecting electrons from the fuel **(1)** and needs to be discharged/connected to an earth line **(1)** otherwise the electrons/static charge could cause a spark. **(1)**
(b) The pipe is connected to an earth line to allow the electrons to flow to earth **(1)** so that there is no risk of a build-up of charge. **(1)**

2 (a) Opposite charges mean the paint droplets are attracted to the object being painted/ an even coverage of paint is obtained. **(1)**
(b) Any two of: minimises wasted paint **(1)**, reduces the amount of paint in the room/ environment **(1)**, reduces paint in the air so it is safer for the painter **(1)**, creates a more even paint coverage. **(1)**

3 (a) Insecticide chemicals can be pumped through the applicator where the chemicals gain electrons **(1)** causing droplets of chemicals to repel and produce a fine mist. **(1)**
(b) The fine mist gives greater coverage to the plants so there is less waste **(1)** and more insects are killed. **(1)**

4 The smoke particles are given a negative charge as they pass through the negatively metal grid **(1)**. They are then attracted to the oppositely / positively charged collecting plates **(1)**. The collecting plates are then shaken to remove the smoke particles. **(1)**

96 Electric fields

1 positive: a positive charge would move outwards **(1)**, field acts radially outwards **(1)**; negative: field acts radially inwards **(1)**, a positive charge would move inwards **(1)**

2 An electric field is a region in space **(1)** where a charged particle may experience a force. **(1)**

3 (a) Diagram showing high-density lines of flux **(1)**, radial field **(1)**, arrows pointing outwards. **(1)**

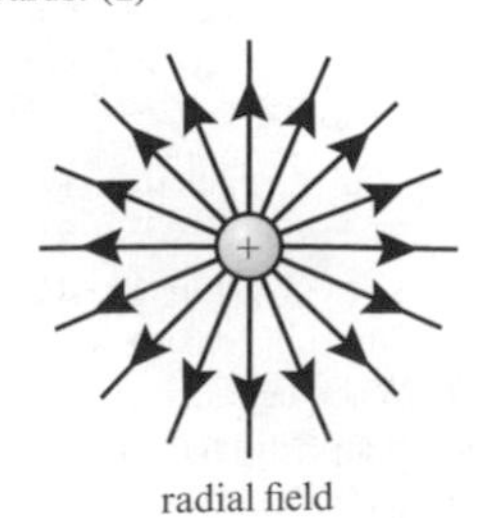

radial field

(b) diagram showing high-density lines of flux **(1)**; parallel field **(1)**; arrows from positive to negative **(1)**

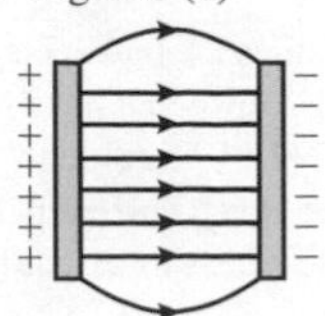

4 (a) (i) outwards **(1)**
(ii) outwards **(1)**
(iii) inwards **(1)**
(b) Repulsion or attraction due to charge will result in a force **(1)** which accelerates the particle. **(1)**

5 The student is right **(1)** because the electrically charged insulator will either be negative or positive due to the build-up/loss of electrons **(1)** (as the electrons will not flow through it) and this acts as a point source creating an electric field. **(1)**

97 Extended response – Static electricity

*Answer could include the following points.

Answers should address any two of the following appliances:

- Photocopiers – points should include:
  - image is projected onto plates
  - light causes the charge to leak away
  - negatively charged toner particles are attracted to the remaining positive charges (toner particles are transferred to paper which is heated).
- Electrostatic precipitators – points should include:
  - smoke particle given a negative charge by passing through a negative grid
  - collecting plates are given a positive charge
  - smoke particles are attracted to the collecting plates (collecting plates are shaken to collect smoke particles).
- Paint spraying – points should include:
  - spray gun is positively charged, which charges the paint
  - similar positive paint particles repel each other and spread out
  - the object to be painted is negatively charged, attracting the paint.
- Insecticide spraying – points should include:
  - insecticide is given a static charge as it leaves the aircraft
  - similar particles repel each other and spread out
  - particles are attracted to the crops in the earth.

98 Magnets and magnetic fields

1 (a) bar magnet: field line out (arrows) at N, **(1)** field lines in (arrows) at S, **(1)** field line close at poles, **(1)** further apart at sides **(1)**

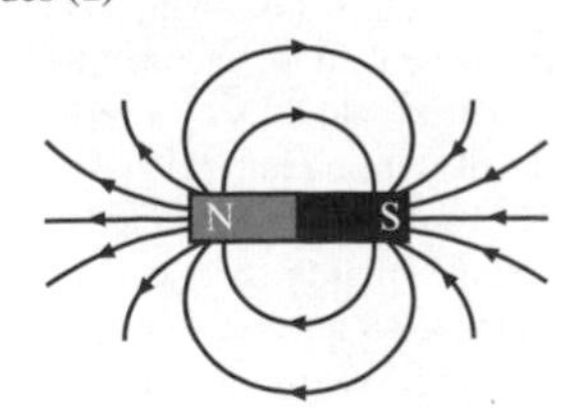

(b) uniform field: parallel field lines, **(1)** arrows from N to S **(1)**

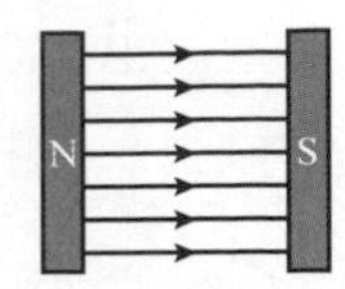

2 Both a bar magnet and the Earth have north and south poles **(1)**. They also both have similar magnetic field patterns. **(1)**

3 A temporary magnet is used for an electric doorbell because it can be magnetised when the current is switched on **(1)**, which attracts the soft iron armature to ring the bell **(1)**, and de-magnetised when the current is switched off **(1)** (returning the armature away from the bell).

4 Rajesh can do a second test by moving a permanent magnet near the magnetic materials **(1)**. Those that are attracted but not repelled will be temporary magnets **(1)**. The materials that can be attracted and repelled are permanent magnets. **(1)**

99 Current and magnetism

1 (a) at least two concentric circles on each diagram, **(2)**
(b) clockwise arrows on cross diagram, **(1)** anticlockwise arrows on dot diagram **(1)**

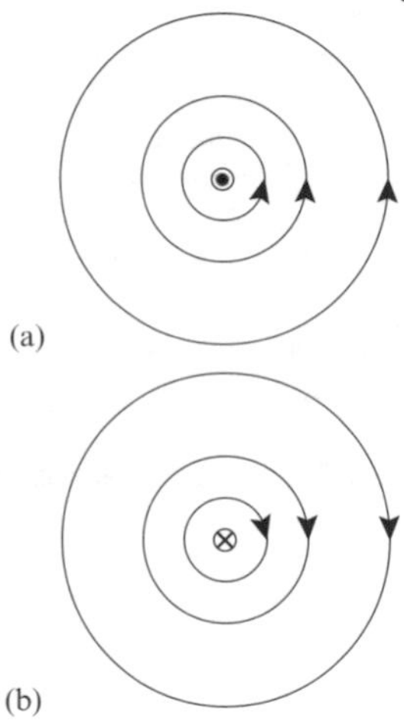

2 B **(1)**

3 (a) The strength of the magnetic field depends on the size of the current in the wire **(1)** and the distance from the wire. **(1)**
(b) (i) x-axis marked 'current' **(1)**;
(ii) x-axis marked 'distance' **(1)**

4 (a) (i) $2B$ **(1)**
(ii) $0.5B$ **(1)**
(b) (i) $0.5B$ **(1)**
(ii) $2B$ **(1)**

100 Current, magnetism and force

1 D **(1)**

2 **F**irst finger: field **(1)**, se**C**ond finger: current **(1)**, thu**M**b: movement **(1)**

3 The size of the force can be increased by increasing the strength of the magnetic field **(1)** or by increasing the current. **(1)**

4 $F = B \times I \times L = 0.0005\ \text{T} \times 1.4\ \text{A} \times 0.30\ \text{m}$ **(1)** $= 0.00021\ \text{N}/2.1 \times 10^{-4}$ **(1)** N **(1)**

5 The interacting magnetic fields result in a force that repels / attracts the coil **(1)**.

The coil then rotates (1). The split ring / commutator swaps connections to the power supply every half turn, reversing the current (1). This reverses the magnetic field (every half turn) creating alternating forces on the coil to keep it spinning. (1)

101 Extended response – Magnetism and the motor effect

*Answer could include the following points.

- A long straight conductor could be connected to a cell, an ammeter and a small resistor to prevent overheating in the conductor.
- When the current is switched on the direction of the magnetic field generated around a long straight conductor can be found using the right-hand grip rule.
- The right-hand grip rule points the thumb in the direction of conventional current and the direction of the fingers show the direction of the magnetic field.
- A card can be cut halfway through and placed at right angles to the long straight conductor. A plotting compass can be used to show the shape and direction of the magnetic field.
- The shape of the magnetic field around the long straight conductor will be circular/concentric circles as the current flows through it.
- The strength of the magnetic field depends on the distance from the conductor.
- The concentric magnetic field lines mean that the field becomes weaker with increasing distance.
- The strength of the magnetic field can be increased by increasing the current.

102 Electromagnetic induction

1 (a) Move the wire up (or down) through a magnetic field. (1)
(b) Move the wire in the opposite direction to (a) OR turn the magnet around but move the wire in the same direction as before. (1)
(c) Any three from: move the wire faster (1), use a stronger magnet (1), use thicker wire/more lengths of wire (1), use more loops/turns in the wire (1), wind the wire around an iron core. (1)

2 The magnets used in power station generators provide a large magnetic field (1) and so they induce a large voltage for transmission (1), reducing current and therefore energy losses due to heating. (1)

3 (a) sine wave sketched that goes above and below the *x*-axis (1)
(b) The current moves between a positive value (1) and a negative value. (1)

103 Microphones and loudspeakers

1 A (1)

2 The cone is moved by a varying force (1) as a result of the interaction of magnetic fields in the speaker (1). The resulting vibration of the cone (1) pushes the air, producing sound waves. (1)

3 The varying potential difference in the wire causes a varying current (1). This produces a varying magnetic field in the wire/coil (1). The interaction of the varying magnetic field in the coil and the permanent magnet (1) results in a varying force exerted on the coil. (1)

4 The compressions/high-pressure areas push the microphone diaphragm in (1) and rarefactions/low-pressure areas cause the diaphragm to move outwards (1). This moves the coil over the magnet (1), which induces a varying potential difference across the ends of the wire. (1)

104 Transformers

1 A step-up transformer is used to increase voltage and decrease current (1). A step-down transformer is used to decrease the voltage and increase the current. (1)

2 (a) The current is reduced. (1)
(b) The lower current reduces the heating effect (1) and therefore less energy is wasted in transmission of electricity. (1)

3 $V_p \div V_s = n_p \div n_s$; $V_p = 230$ V, $V_s = 19$ V, $n_s = 380$ turns; $n_p = V_p \times n_s \div V_s$ (1) = 230 V × 380 turns ÷ 19 V (1) = 4600 turns (1)

4 (a) This is a step-down transformer (1) because there are more turns on the primary coil than on the secondary coil. (1)
(b) $n_p = 600$ turns, $n_s = 20$ turns; $V_p = 360$ V; $V_s = V_p \times n_s \div n_p$ (1) = 360 V × 20 turns ÷ 600 turns (1) = 12 V (1)

105 Transmitting electricity

1 (a) As the voltage is increased, the current goes down (1) so this reduces the heating effect due to resistance (1) and means less energy is wasted in transmission. (1)
(b) The voltages are high enough to kill you if you touch or come into contact with a transmission line. (1)

2 step-down transformer: decreases voltage (1)

3 $P = I \times V$ = 20000 A × 25000 V (1) = 500000 kW (or 500 MW) (1)

4 Any two from: Step-up transformers increase voltage and so lower the current, reducing the heating losses (1). Wires are thermally insulated (1). Wires of low resistance are used. (1)

5 Step-up transformers are used to increase the voltage (1) as it leaves the power station for transmission through the National Grid (1). Near homes, step-down transformers are used to reduce the voltage (1) to make it safer for use in houses. (1)

106 Extended response – Electromagnetic induction

*Answer could include the following points.

- Microphone:
  - Sound waves move the diaphragm which is connected to a coil/solenoid.
  - The movement of the coil/solenoid around a permanent magnet induces a changing current/signal in the coil.
  - The chan ging current/signal induced in the coil is at the same frequency as the sound waves.
  - The changing current/signal is sent to an amplifier.
- Headphones:
  - The wire from the amplifier carries a changing current.
  - The changing current, which has its own magnetic field, moves through a coil/solenoid around a permanent magnet.
  - The magnetic field of the changing current interacts with the field of the permanent magnet, causing movement.
  - The movement of the coil/solenoid moves a diaphragm which pushes air, creating sound waves.

107 Changes of state

1 All four links correct for **2 marks**, three correct for **1 mark**, fewer than three correct for **0 marks**. particles move around each other – liquid – some intermolecular forces; particles cannot move freely – solid – strong intermolecular forces; particles move randomly – gas – almost no intermolecular forces

2 (a) They all consist of particles. (1)
(b) Particles have different amounts in the kinetic energy store (1) and experience different intermolecular forces. (1)

3 B (1)

4 At boiling point the liquid changes state (1) so the energy applied after boiling point is reached goes into breaking bonds (1) between the liquid particles. The particles gain more energy and become a gas. (1)

5 The kinetic energy (1) of the particles decreases (1) as the ice continues to lose energy to the surroundings. This is measured as a fall in temperature. (1)

108 Density

1 $\rho = m \div V$ (1) = 4000 g ÷ 5000 cm^3 (1) = 0.8 g/cm^3 (1)

2 A, (1) B (1)

3 volume = 10 cm × 25 cm × 15 cm = 3750 cm^3 (1); $\rho = m \div V$ so $m = \rho \times V$ = 3 g/cm^3 × 3750 cm^3 (1) = 11 250 g (1) = 11.25 kg (1)

4 Marco has approached this problem by stating a scientific principle (1) relating density to states of matter but he has not tried to investigate this (1). Ella has approached this problem by observing and comparing (1) different densities, but she has not tried to explain this (1). Both students should expand their approach so that observations are explained through scientific principles. (1)

109 Investigating density

1 (a) mass (1)
(b) electronic balance (1)

2 (a) The volume of mass may be found by measuring its dimensions (1) or by using a displacement method, such as immersing the object in water in a measuring cylinder, to measure how much liquid the mass displaces. (1)
(b) The measurement method is suitable for regular-shaped objects (1) whereas the displacement method is best for irregular-shaped objects, where measuring dimensions would be more difficult. (1)

3 (a) Place a measuring cylinder on a balance and then zero the scales with no liquid present in the measuring cylinder (1). Add

the liquid to the required level **(1)**. Record the mass of the liquid (in g) from the balance and its volume (in cm^3) from the measuring cylinder. **(1)**

(b) Take the value at the bottom of the meniscus **(1)**. Make sure the reading is taken with the line of sight from the eye to the meniscus perpendicular to the scale to avoid a parallax error. **(1)**

(c) density = mass ÷ volume = 121 g ÷ 205 cm^3 = 0.59 **(1)** g/cm^3 **(1)**

(d) Any one from: Care should be taken when placing liquid on an electronic balance to avoid the risk of electrical shock from wet hands or spillage when using electricity **(1)** – keep hands and working area dry **(1)**. Care should be taken to avoid spillages to avoid the risk of slippage **(1)** – wipe spills and warn others **(1)**. Do not use toxic or harmful liquids **(1)** – always check the hazard label on a liquid container. **(1)**

110 Energy and changes of state

1 Specific heat capacity is a measure of how much energy is required to change the temperature of a mass of 1 kg by 1 °C. **(1)**

2 $\Delta Q = m \times c \times \Delta T$ **(1)** = 0.8 kg × 4200 J/kg °C × 50 °C **(1)** = 168 000 J **(1)**

3 $Q = m \times L$ = 25 kg × 336 000 J/kg **(1)** = 8 400 000 J **(1)**

4 (a) melting: added to lower horizontal line; boiling: added to higher horizontal line **(1)** (both needed for the mark)

(b) The energy being transferred to the material is breaking bonds **(1)** and as a result the material undergoes a phase change. **(1)**

5 $c = \Delta Q \div (m \times \Delta T)$ so $\Delta Q_{in} = I \times V \times t$ = 2.4 A × 12 V × (9 × 60) s **(1)** = 15 552 J **(1)**; $\Delta Q_{in} = \Delta Q_{out}$ so c = 15 552 ÷ (0.8 kg × 25 °C) **(1)** = 15 552 ÷ 20 = 777.6 J/kg °C **(1)**

111 Thermal properties of water

1 (a) the amount of energy required to raise the temperature of 1 kg of material by 1 K (or 1 °C) **(1)**

(b) specific heat capacity = change in thermal energy ÷ (mass × change in temperature) or ($c = \Delta E \div (m \times \Delta T)$ **(1)**

2 (a) Place a beaker on a balance, zero the balance and add a measured mass of water **(1)**. Take a start reading of the temperature **(1)**. Place the electrical heater into the water and switch on **(1)**. Take a temperature reading every 30 seconds **(1)** until the water reaches the required temperature. **(1)**

(b) Measure the current supplied, the potential difference across the heater and the time for which the current is switched on **(1)**. Use these values to calculate the thermal energy supplied using the equation $E = V \times I \times t$. **(1)**

(c) Add insulation around the beaker **(1)** so less thermal energy is transferred to the surroundings and a more accurate value for the specific heat capacity of the water may be obtained. **(1)**

3 Plot a graph of temperature against time **(1)**. The changes of state are shown when the graph is horizontal (the temperature is not increasing). **(1)**

4 Both experiments use an electrical heater close to water so there is a danger of electric shock – keep electrical wires and switches dry **(1)**. Both experiments use water that could be spilled and cause slippage – report and wipe up immediately. **(1)** (Note: specific latent heat experiments tend not to use glass beakers (which could break and cause cuts in the specific heat capacity experiment) but tend to use metal containers, so glass is not necessarily common to both experiments. The specific heat capacity experiment does not require water to be heated to a level to cause scalds so the hot water/water vapour hazard in the specific latent heat experiment is not necessarily common to both experiments.)

112 Pressure and temperature

1 Temperature is a measurement of the average kinetic energy of the particles in a material. **(1)**

2 (a) 273 K → 0 °C, **(1)** 255 K→ –18 °C, **(1)** 373 K→ 100 °C **(1)**

(b) (i) At absolute zero, the volume/pressure **(1)** and kinetic energy of the particles **(1)** of a substance will be zero. **(1)**

(ii) –273 °C **(1)**

3 (a) As the temperature increases the particles will move faster **(1)** because they gain more energy. **(1)**

(b) As the particles are moving faster they will collide with the container walls more often **(1)** therefore increasing the pressure. **(1)**

(c) It increases. **(1)**

4 The kinetic energy of the particle will also increase by a factor of four **(1)** because temperature and average kinetic energy are directly proportional. **(1)**

113 Volume and pressure

1 When particles of a gas collide with a surface **(1)** they exert a force at right angles to the surface **(1)** resulting in pressure. **(1)**

2 C **(1)**

3 $P_1 \times V_1 = P_2 \times V_2$ so $V_2 = P_1 \times V_1 \div P_2$ **(1)** so V_2 = 100 kPa × 230 cm^3 ÷ 280 kPa **(1)** = 82.2 **(1)** cm^3

4 (a) $P_1 = (P_2 \times V_2) \div V_1$ **(1)** = (640 litres × 100 kPa) ÷ 8 litres **(1)** = 8000 kPa (or 8 MPa, 8 × 10^6 Pa) **(1)**

(b) time = total volume ÷ rate so time = 640 litres ÷ 2 litres/min = 320 min **(1)**, but 8 litres will be left in the cylinder at atmospheric pressure so gas will only last 316 min **(1)**

114 Extended response – Particle model

*Answer could include the following points.

Thermal energy transfer reduced in each area:

- Roof: loft insulation traps air (insulator) reducing thermal energy transfer by conduction (vibration of solid particles) through room ceiling.
- Walls: cavity wall insulation (insulating) foam traps air (insulator) reducing thermal energy transfer by conduction (vibration of solid particles) through walls.
- Floor: carpets (insulating material) trap air (insulator) reducing thermal energy transfer by conduction (vibration of solid particles) through floor.
- Doors: fit draught excluders to reduce the cooling effect from draughts reducing thermal energy transfer by convection (movement of fluid particles) around room.
- Windows: the space between the panes of glass in double (or triple) glazing reduces thermal energy transfer by conduction (vibration of solid particles) through glass.
- Windows: heavy/thick curtains (insulating material) reduce thermal energy transfer by conduction (vibration of solid particles) through glass.

115 Elastic and inelastic distortion

1 push forces (towards each other): compression **(1)**, pull forces (away from each other): stretching **(1)**, clockwise and anticlockwise: bending **(1)**

2 (a) washing line (or any valid example) **(1)**

(b) G-clamp, pliers (or any valid example) **(1)**

(c) fishing rod (with a fish on the line) (or any valid example) **(1)**

(d) dented can or deformed spring (or any valid example) **(1)**

3 After testing, beam 1 would return to the same size and shape as prior to the test **(1)** under the load but would be intact **(1)**. Beam 2 would distort and change shape **(1)** but would (probably) still be intact. **(1)**

4 Car manufacturers install crumple zones/seat belts/airbags **(1)** in cars. These are parts of the car body that are designed to distort/change shape **(1)** in the event of a crash. They extend the time taken for a body to come to rest, reducing the force on the body. **(1)**

116 Springs

1 Elastic means that the object will return to original size and shape **(1)** (*both needed for mark*) after the deforming force is removed. **(1)**

2 extension = 0.07 m – 0.03 m = 0.04 m; force = spring constant/k × extension = 80 N × 0.04 m **(1)** = 3.2 **(1)** N **(1)**

3 D **(1)**

4 (a) $F = k \times x$ so $k = F \div x$ **(1)** = 30 N ÷ 0.15 m **(1)** = 200 N/m **(1)**

(b) $E = \frac{1}{2} \times k \times x^2 = \frac{1}{2}$ × 200 N/m × (0.15 m)^2 **(1)** = 2.25 J **(1)**

117 Force and springs

1 (a) Hang a spring from a clamp attached to a retort stand and measure the length before any masses or weights are added using a half-metre ruler, marked in mm **(1)**. Carefully add the first mass or weight and measure the total length of the extended spring **(1)**. Unload the mass or weight and re-measure the spring to make sure that the original length has not changed **(1)**. Add at least five masses or weights and repeat the measurements each time. **(1)**

(b) The elastic potential energy can all be recovered **(1)** and is not transferred to cause a permanent change of shape in the spring. **(1)**

(c) Masses must be converted to force (N) by using $W = m \times g$ / $F = m \times g$ **(1)**. The extension of the spring must be calculated for each force by taking away the original length of the spring from each reading **(1)**. Extension measurements should be converted to metres. **(1)**

(d) (i) The area under the graph equals the work done/the energy stored in the spring as elastic potential energy. **(1)**

(ii) The gradient of the linear part of the force–extension graph gives the spring constant k. **(1)**

(e) Hooke's law **(1)**

(f) energy stored = $\frac{1}{2} \times k \times x^2$ **(1)**

2 The length of a spring is measured with no force applied to the spring whereas the extension of a spring is the length of the spring measured under load/force less the original length. **(1)**

118 Pressure and fluids

1 A column of air reaching from the Earth's surface to the top of the atmosphere, covering one square metre of the Earth **(1)**, containing a mass of air of 10000 kg with a weight of 100000 N **(1)**

2 Water pressure increases with depth because density increases with depth **(1)** so more particles create greater pressure. **(1)**

3 $P = h \times \rho \times g$ = 1500 m × 1025 kg/m^3 × 10 N/kg **(1)** = 15375000 Pa **(1)**

4 area of football pitch = 100 m × 50 m **(1)** = 5000 m^2; atmospheric pressure = 100000 Pa **(1)**; total force on pitch = $P \times A$ = 100000 Pa × 5000 m^2 **(1)** = 5×10^8 N **(1)**

119 Upthrust and pressure

1 B, **(1)** D **(1)**

2 1000 N **(1)**

3 When the ship is fully loaded it has a greater weight **(1)** which displaces more water. **(1)**

4 (a) $P = F \div A$ = 350 N ÷ 0.3 m^2 **(1)** = 1167 Pa **(1)**

(b) $A = F \div P$ = 15 N ÷ 120 Pa **(1)** = 0.125 m^2 **(1)**

(c) $P = (m \times g) \div A$ **(1)** = (6 kg × 10 N/kg) ÷ 0.25 m^2 **(1)** = 240 Pa **(1)**

5 The stiletto heel has a smaller surface area than the slipper heel **(1)**. This means that the force per area of the stiletto results in a much greater pressure **(1)** than the force per area of the slipper, which exerts a lower pressure **(1)**. The larger pressure of the stiletto heel causes greater damage to the old wooden floors. **(1)**

120 Extended response – Forces and matter

*Answer could include the following points.

- The submarine will float when its weight is equal to the weight of the water it displaces.
- The weight of water displaced is equal to the upthrust to keep the submarine floating.
- The upwards force is called the buoyancy force.
- To overcome the buoyancy force the submarine can become denser than the water displaced by taking on more water, so the submarine sinks.
- When the upthrust is greater than the weight, the weight of water displaced is greater than the weight of the submarine so it rises when it is underwater.
- The submarine will sink if its density is greater than that of the surrounding water.
- The submarine will sink if it displaces a weight of water that is less than its own weight.
- When the submarine resurfaces, water is pumped out of the submarine, reducing its density, compared to the surrounding water, causing it to rise.

121 Timed Test 1

1 (a) Measure, mark and record the distance to be travelled by the trolley **(1)**. Place the trolley at the start/top of the ramp and release **(1)**. Time how long it takes to cover the marked distance **(1)** (both distance and timing should be mentioned for the mark). Repeat the experiment to reduce the influence of random errors. **(1)**

(b) Any two from: light gates **(1)**, data logger **(1)**, computer **(1)**

(c) (i) distance, time; independent variable = height **(1)**; (all three needed for mark)

(ii) speed (m/s) **(1)** and any two from: 0.56, 0.75, 1.0, 1.29, 1.8, 4.5, 9.0 **(1)**

(d) The forces of friction and air resistance would both oppose the force of the trolley **(1)**. There would be friction due to the contact between the wheels and the surface/floor **(1)** and the axles/wheels **(1)**. The opposing forces would continue to act opposite to the force of the trolley until the resultant force becomes zero. **(1)**

2 (a) $a = (v - u) \div t$ **(1)**

(b) average speed = 5 m/s so d = 5 m/s × 20 s **(1)** = 100 m **(1)**

(c) momentum = 1000 kg × 10 m/s **(1)** = 10000 **(1)** kg m/s **(1)**

(d) C **(1)**

3 (a) (i) background radiation **(1)**

(ii) Any two from: radon gas **(1)**, cosmic rays **(1)**, medical uses **(1)**, nuclear industry **(1)**, natural sources/rocks. **(1)**

(iii) Several readings are taken to identify anomalies to gain a reading close to the true value. **(1)**

(b) (i) y-axis labelled 'Corrected count rate (counts per minute)' and x-axis labelled 'Time (minutes)' **(1)**, all points correctly plotted (+/– half a square) **(2)** or six or more points correctly plotted (+/– half a square) **(1)**

(ii) single exponential decay curve of best fit passing through six points **(1)**

(iii) 2 min **(1)**

(c) Alpha radiation is highly ionising because it is the most massive particle so it can easily ionise atoms by knocking off their electrons **(1)**. Beta radiation is moderately ionising: the particles are highly energised although they are very small so they have less chance of knocking electrons off other atoms **(1)**. Gamma radiation is the least ionising because it has no mass but the gamma waves may still ionise other atoms. **(1)**

4 (a) C **(1)**

(b) Gravity causes nebulous dust and gases to move together which eventually form a star **(1)**. When the star is in main sequence, gravity is balanced by the outward force of the expanding hot gases and the star is stable **(1)**. When these forces become unbalanced, gravity is less because there is less mass **(1)** and the star expands. Following the red giant phase, gravity causes the remaining matter to condense to form a white dwarf. **(1)**

5 (a) (i) A should be marked inside the first half wavelength between the time axis and a crest/trough **(1)**

(ii) λ should be marked between the start and end of an identified single wave **(1)**

(iii) wave speed = frequency × wavelength or $v = f \times \lambda$ **(1)**

(iv) T = 0.0001 / 2
T = 0.00005s **(1)**

(v) 2 wavelengths = 0.0001 s so period = 0.00005 s, frequency = 1 ÷ period = 1 ÷ 0.00005 s **(1)** = 20000 Hz **(1)**

(b) (i) sonar **(1)**

(ii) The ship emits sound waves which travel down to the shoal of fish. **(1)** Detectors on the ship receive the echo of the sound waves as they are reflected back from the fish. **(1)** The depth/location of the fish can be found by calculating the time between the sound wave being sent and the echo being detected, and dividing this by two. **(1)**

6 (a) (i) microwaves: communication/cooking **(1)**

(ii) ultraviolet: security marking/disinfecting water **(1)**

(iii) gamma rays: detecting cancer/treating cancer/sterilising food/water **(1)**

(b) Gamma rays **(1)** are the most damaging to body cells because they carry the most energy and are highly ionising radiation. **(1)**

(c) D **(1)**

(d) $v = x \div t$ so $x = v \times t = 3 \times 10^8$ m/s × 500 s **(1)** = 1.5×10^{11} m **(1)** = 1.5×10^8 km **(1)**

(e) Satellites are not used to transmit radio waves because they are outside the Earth's atmosphere and radio waves cannot pass through the ionosphere **(1)**. They are reflected back towards the Earth by the ionosphere **(1)**. Satellites are used to transmit microwaves back to a receiving aerial on Earth from beams (using dish antennae) which are sent to a satellite **(1)**, but microwaves can penetrate the ionosphere. **(1)**

7 (a) total internal reflection **(1)**

(b) (i) i marked between incident ray and normal **(1)**, r marked between normal and refracted ray (inside the block) **(1)**

(ii) As the wave passes from a less dense medium (air) to a denser medium (glass) the speed of light slows down **(1)** as the leading edge of the wave front reaches the glass first. The wave then changes direction as the whole wave pivots on this leading edge. **(1)**

(c) Rays drawn: tip of object to mid-lens then down through F_2, **(1)** second ray from tip of object through centre of lens to meet first ray **(1)**, position of image shown. **(1)**

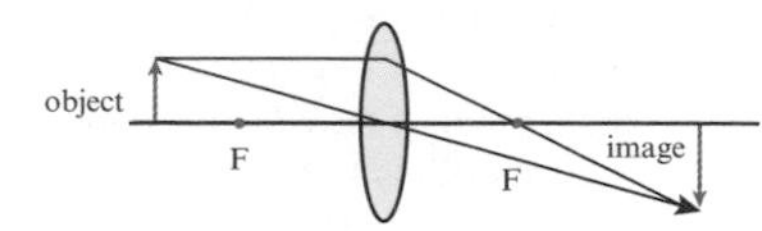

8 (a) (i) $\Delta GPE = m \times g \times h = 750$ kg × 10 N/kg × 15 m **(1)** = 112 500 J **(1)**
(ii) $P = \Delta GPE \div t = 112\,500$ J ÷ 20 s **(1)** = 5625 **(1)** W **(1)**
(b) thermal energy store of the motor **(1)**, thermal energy store of the elevator materials. **(1)**
(c) $v = d \div t = 15$ m ÷ 20 s **(1)** = 0.75 m/s **(1)**; KE = ½ × $m \times v^2$ = ½ × 750 kg × $(0.75 \text{ m/s})^2$ **(1)** = 210.9 J **(1)**
(d) This process could be described as wasteful because it causes a rise in temperature in parts of the system, so dissipating energy in heating the surroundings **(1)**. A rise in temperature happens in (any one of): the lift motor **(1)**, the fabric of the lift **(1)**, the lift cables. **(1)**

9 (a) B **(1)**
(b) (i) The coloured filters for red, blue and green only allow one colour through **(1)** and block all the others. **(1)**
(ii) trousers: blue **(1)**, shirt: black **(1)**
(d) total weight of spotlights = 900 N and total area = 4.5 m² **(1)**; $P = F \div A$ = 900 N ÷ 4.5 m² **(1)** = 200 Pa **(1)**

10 (a) Earth, Uranus and Neptune **(1)** (all three needed for mark)
(b) (i) A natural satellite is a naturally occurring body that orbits a larger body. **(1)**
(ii) any example of a natural satellite, e.g. a moon **(1)**
(c) The circular path means a continual change of direction **(1)** and direction is a component of velocity. **(1)**
*(d) *The points below are not a prescriptive answer and it is not required to include all the material which is indicated as relevant. Additional content included in the response must be scientific and relevant. Answers may include the following:*
As emitters of light move away from us the observed wavelength of light reaching Earth is longer/observed to be in red-shift **(1)** when compared to observations of light made on Earth **(1)**. We can see this from the absorption spectra analyses which show the black absorption lines **(1)** all shifted towards the red end of the spectrum **(1)**. This observation supports the theory of the expanding Universe because the objects that are travelling the fastest have greater red-shift **(1)** and are therefore farthest away. **(1)**

125 Timed Test 2

1 (a) (i) $P = V \times I$ or power = voltage × current **(1)**
(ii) $I = P \div V$ = 2000 W ÷ 230 V **(1)** = 8.7 A **(1)**
(iii) A fuse with too high a rating could result in too much current and the risk of a fire **(1)**. Too low a rating would melt and break the circuit each time the kettle was switched on. **(1)**
(iv) C **(1)**
(b) (i) E = Q × V. **(1)**
(ii) Current is the rate of flow of electric charge. **(1)**
(iii) Coulombs **(1)**
(c) Energy is transferred to the thermal energy store of the kettle **(1)**. The heater/element of the kettle transfers energy to the thermal energy store of the water and the kettle jug **(1)**. The water (as steam) and the kettle transfer energy to the thermal energy store of the environment. **(1)**

2 (a) The direction of the magnetic field will be clockwise **(1)** as the current flows from positive to negative **(1)** (down to up) and the right-hand screw rule **(1)** shows the field to be clockwise.
(b) (i)

magnetic field
coil carrying electric current

correct shape **(1)**; lines of flux very close inside **(1)**; lines of flux further apart outside **(1)**; pattern indicated on both sides of electromagnet **(1)**
(ii) any valid use, e.g. doorbell, speaker, electromagnet. **(1)**
(c) $F = B \times I \times l$ = 0.5 T × 3 A × 0.75 m **(1)** = 1.1 **(1)** N **(1)**

3 (a) $E = I \times V \times t$ = 12 A × 12 V × 120 s **(1)** = 17 280 **(1)** J **(1)**
(b) ΔQ $(E) = m \times c \times \Delta T$ so $\Delta T = \Delta Q \div (m \times c)$ = 17 280 J ÷ (2 kg × 385 J/kg K) **(1)** = 22.4 **(1)** K/°C **(1)**
(c) Some thermal energy is dissipated to the environment. **(1)**
(d) (i) Unwanted energy transfer can be reduced by thermally insulating the block. **(1)**
(ii) An example could be a non-flammable material that has air pockets as air is a good insulator/poor conductor. **(1)**

4 (a) (i) The metal rod is a conductor **(1)** and so discharges the static electricity to earth. **(1)**
(ii) The static charge on the helicopter could cause a spark to ignite fuel vapours nearby **(1)** or cause a shock to persons who step out of the aircraft. **(1)**
(b) As the fuel flows through the pipe from the tanker to the plane **(1)**, a static charge could build up on the fuelling line due to friction **(1)**. The fuelling line must be earthed to conduct the electrons to earth to prevent a spark. **(1)**
(c)* *The points below are not a prescriptive answer and it is not required to include all the material which is indicated as relevant. Additional content included in the response must be scientific and relevant. Answers may include the following:*
In a thundercloud, huge convection currents occur **(1)** which move around the cloud particles/water and ice **(1)**. These are forced up by warm air and down by gravity **(1)**. The cloud particles strip negative electrons from the air/ice **(1)** which build up **(1)** at the base of the cloud. This is because the cloud is insulated **(1)** from the earth by the air. When the negative charge of the cloud becomes large enough **(1)**, the electrons discharge to earth in the form of lightning. **(1)**

5 (a) Each lamp would have a potential difference of 0.5 V **(1)** as potential difference is shared in a series circuit. **(1)**
(b) voltmeter connected in parallel across one of the lamps **(1)**
(c) As current is shared in a series circuit you could add another cell **(1)** or you could reconnect the circuit to become a parallel circuit. **(1)**
(d) (i) The resistance of a thermistor changes with temperature **(1)**. When the temperature rises, more electrons are released **(1)** causing more current to flow and reducing resistance. **(1)** (Accept converse.)
(ii) It could be used in a temperature-sensing circuit. **(1)**

6 (a) All three correct for **2 marks**, two correct for **1 mark**. Solid: closely packed particles in fixed positions within a regular lattice. Liquid: closely packed particles, irregular arrangement. Gas: few particles, spread out in random arrangement.
(b)* *The points below are not a prescriptive answer and it is not required to include all the material which is indicated as relevant. Additional content included in the response must be scientific and relevant. Answers may include the following:*
Particles in a solid have a relatively low amount of kinetic energy compared with the other states of matter **(1)** and remain in molecular bonding with neighbouring atoms **(1)**. Particles in a liquid have higher amounts of kinetic energy than solids **(1)** but lower amounts of kinetic energy than gases **(1)** and remain in weak molecular bonding with/are able to move around neighbouring atoms **(1)**. Particles in a gas have high amounts of kinetic energy **(1)** and are able to move independently of other neighbouring atoms. **(1)**

7 (a) density = mass ÷ volume or $\rho = m \div V$ **(1)**
(b) $\rho_1 = m \div V$ = 1.5 kg ÷ 0.2 m³ = 7.5 kg/m³ **(1)**; ρ_2 = 0.7 kg ÷ 0.15 m³ = 4.7 kg/m³ **(1)**; ρ_3 = 0.7 kg ÷ 0.2 m³ = 3.5 kg/m³ **(1)** so block 1 has the highest density **(1)**
(c) (i) The student must make sure that the line of sight from the eye to the meniscus **(1)** is perpendicular to the scale **(1)** (to avoid parallax errors).
(ii) The student must measure the mass of the rock **(1)**. This can be done on a balance or by hanging it from a suspended force meter. **(1)**
(d) The student is right because she used different methods to calculate the volume of the objects **(1)** so this could have affected her calculation of density. **(1)**

8 (a) The step-up transformer increases the potential difference leaving the power station so that the current is reduced **(1)**. This reduces the amount of energy transferred to the thermal store/wasted during the transmission of electricity through the National Grid. **(1)**
(b) $V_p \div V_s = n_p \div n_s$ so $V_s = n_s \times V_p \div n_p$ **(1)** = 4500 × 250 V ÷ 150 **(1)** = 7500 V **(1)**
(c) The high voltages that are necessary for efficient transmission via the National

Grid are too dangerous **(1)** so a step-down transformer is used to reduce the voltage. **(1)**

(d) (i) $V_p \times I_p = V_s \times I_s$ so $I_s = (V_p \times I_p) \div V_s$ **(1)** = (4600 V × 5 A) ÷ 230 V **(1)** = 100 A **(1)**

(ii) Assume that the transformer is 100% efficient. **(1)**

9 (a) elastic limit / limit of proportionality **(1)**

(b) Any two of the following; add smaller masses each time **(1)**; unload the spring after each mass is added to check spring has not stretched **(1)**; extend using a second variable – different size spring / different material, **(1)**; any other valid suggestion. **(1)**

(c) 0.1:**36, 0.2**:40, 0.3:**44, 0.5**:52 (all four correct – **2 marks**; three correct – **1 mark**).

(d) In linear elastic deformation, the extension (of a spring) is directly proportional to the force added **(1)**. The graph is a straight line which shows a directly proportional relationship **(1)**. The spring is loaded with weights of 0.1 N at a time **(1)** and the spring extends by 4 mm with each 0.1 N. **(1)**

10 (a) B **(1)**

(b)* *The points below are not a prescriptive answer and it is not required to include all the material which is indicated as relevant. Additional content included in the response must be scientific and relevant. Answers may include the following:*

The air particles have lost kinetic energy overnight **(1)** as the temperature has fallen **(1)**. The particles slow down **(1)** and so they do not collide with the balloon walls so often **(1)**, therefore pressure in the balloon is less than on a hot day **(1)**. When the air particles gain kinetic energy **(1)** due to a rise in temperature **(1)** they move faster **(1)** and collide with the balloon wall more often, resulting in higher pressure pushing the balloon wall outwards. **(1)**

Physics Equations List

(final velocity)^2 – (initial velocity)^2 = 2 × acceleration × distance

$v^2 - u^2 = 2 \times a \times x$

force = change in momentum ÷ time

$F = \frac{(mv - mu)}{t}$

energy transferred = current × potential difference × time

$E = I \times V \times t$

force on a conductor at right angles to a magnetic field carrying a current = magnetic flux density × current × length

$F = B \times I \times l$

$\frac{\text{voltage across primary coil}}{\text{voltage across secondary coil}} = \frac{\text{number of turns in primary coil}}{\text{number of turns in secondary coil}}$

$\frac{V_p}{V_s} = \frac{N_p}{N_s}$

potential difference across primary coil × current in primary coil = potential difference across secondary coil × current in secondary coil

$V_p \times I_p = V_s \times I_s$

change in thermal energy = mass × specific heat capacity × change in temperature

$\Delta Q = m \times c \times \Delta\theta$

thermal energy for a change of state = mass × specific latent heat

$Q = m \times L$

$P_1 V_1 = P_2 V_2$

to calculate pressure or volume for gases of fixed mass at constant temperature

energy transferred in stretching = 0.5 × spring constant × (extension)^2

$E = \frac{1}{2} \times k \times x^2$

pressure due to a column of liquid = height of column × density of liquid × gravitational field strength

$P = h \times \rho \times g$

Your own notes

Your own notes

Your own notes

Published by Pearson Education Limited, 80 Strand, London, WC2R 0RL.

www.pearsonschoolsandfecolleges.co.uk

Copies of official specifications for all Edexcel qualifications may be found on the website: www.edexcel.com

Typeset and produced by Phoenix Photosetting
Illustrated by TechSet Ltd.
Cover illustration by Miriam Sturdee

First published 2017

20 19
10 9 8 7 6 5 4

British Library Cataloguing in Publication Data
A catalogue record for this book is available from the British Library

ISBN 978 1 292 13368 3

Printed in Great Britain by Bell and Bain Ltd, Glasgow

Acknowledgements
The author and publisher would like to thank the following organisations for permission to reproduce photographs and figures:

Photographs
Alamy Images: Damian Gil 27; **NASA:** 9

All other images © Pearson Education

Figures
Figure showing world energy use on page 20, BP Statistical Review of World Energy 2012 www.bp.com/statisticalreview